中国创新设计红星奖年鉴

China Red Star Design Award Yearbook

09

中国创新设计红星奖委员会 主编
China Red Star Design Award Committee

中国建筑工业出版社
China Architecture & Building Press

图书在版编目（CIP）数据

中国创新设计红星奖年鉴（2009）／中国创新设计红星奖委员会主编.—北京：中国建筑工业出版社，2009
ISBN 978-7-112-11593-8

Ⅰ.中… Ⅱ.中… Ⅲ.工业产品－造型设计－中国－2009－年鉴 Ⅳ.TB472-54

中国版本图书CIP数据核字（2009）第207314号

责任编辑：马　彦

装帧设计：肖晋兴

责任校对：袁艳玲 陈晶晶

中国创新设计红星奖年鉴（2009）
2009 China Red Star Design Award Yearbook

中国创新设计红星奖委员会 主编
China Red Star Design Award Committee

*

中国建筑工业出版社出版、发行（北京西郊百万庄）

各地新华书店、建筑书店经销

北京凌奇印刷有限责任公司印刷

*

开本：850×1168毫米　1／12　印张：13⅓　字数：470千字

2009年12月第一版　2009年12月第一次印刷

定价：118.00元

ISBN 978-7-112-11593-8

(18840)

2009中国创新设计红星奖年鉴编委会

2009 China Red Star Design Award Yearbook Editorial Committee

温家宝总理于2007年2月所作的重要批示

"It should attach great importance to industrial design"
—— Comments by Premier Wen Jiabao on Feb.2007.

工业设计

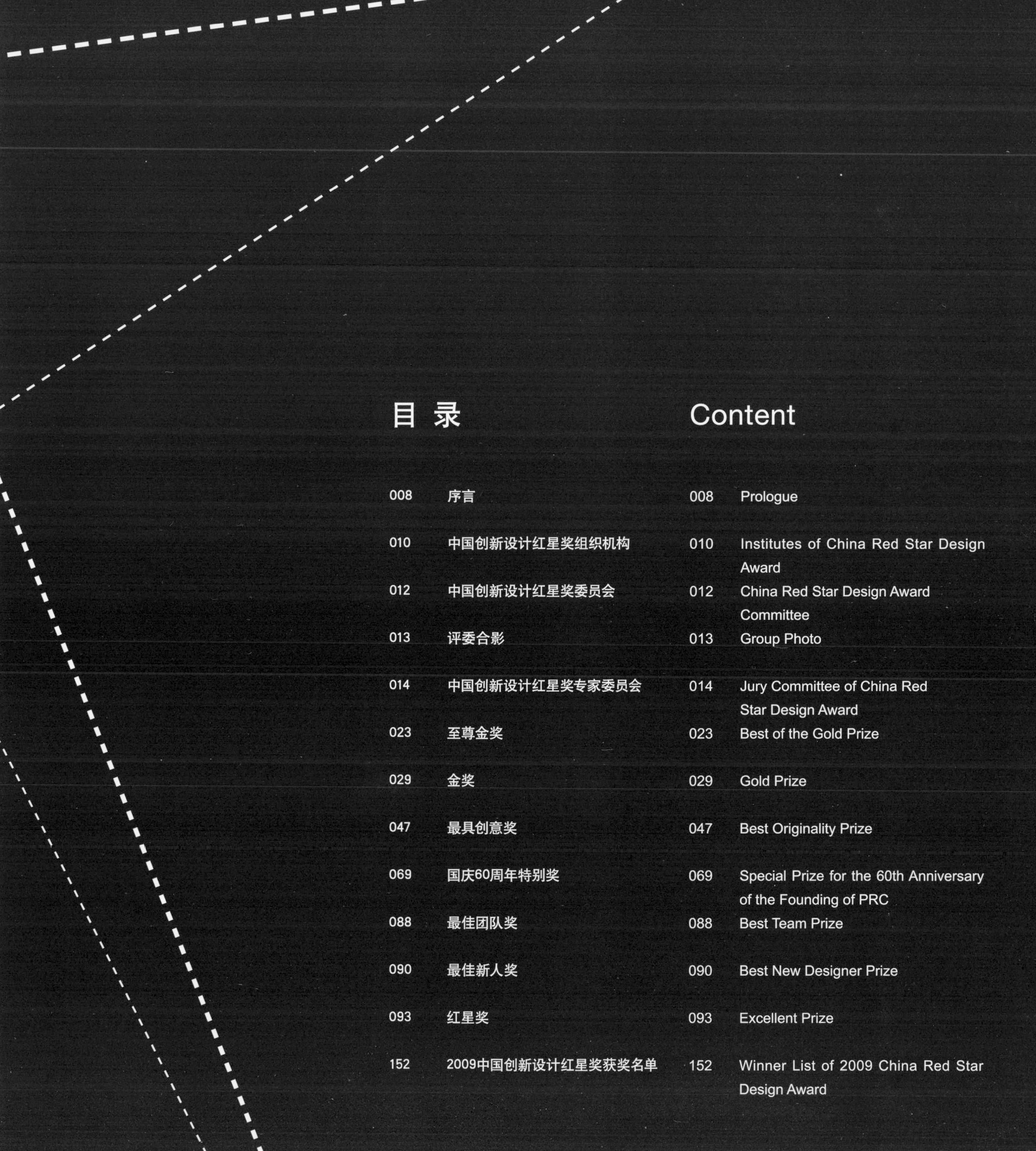

目 录

Content

迎接工业设计的新机遇

2009年是红星奖创立的第四年，也是红星奖不断完善、走向成熟的一年。国际化的专家评委、严格规范的评审程序、广泛的企业参与、高水平的参评产品、隆重的颁奖典礼和国际化的路演推介、Icsid权威互认，使得红星奖成为一个具有广泛公信力的国际化品牌。

2009年，红星奖共征集到758家企业的3821件产品，产品覆盖范围达到19个省、4个市和 2个地区，行政区县涵盖范围达到74%。报名企业数量和产品数量分别比去年增加34%和21%，其中家居用品领域参评产品与去年相比增幅达146%，装备类增幅达100%。参评单位中企业的比例为76%，吸引了联想、海尔、华帝、欧琳、美的、曲美家具、探路者等大企业连续参评，龙域设计、UI design、蒙恬科技、优乐文具等曾荣获红点奖的设计公司也积极参加。

四年来，美国、英国、德国、法国、意大利、日本、韩国、澳大利亚、中国以及中国台湾、中国香港等12个国家和地区的68位专家担任红星奖评委，来自设计、科技、经济、媒体、市场营销等多个领域。红星奖评委国际化、普遍性、跨领域、多学科的特点，保证了红星奖评审的权威性和公正性。

红星奖以提升品牌价值为核心,先后受邀到香港、天津、上海、福州、西安、南京、青岛、宁波、深圳、杭州、成都、广州等18个城市开展28场次巡展；组织国内外设计公司与当地制造业企业进行了20余场次对接活动；为获奖企业在各地路演中组织营销推介活动，获得订单，取得了较好的收益。

随着红星奖影响力的扩大，其社会反响也日益凸现，长虹、创维、康佳、华旗资讯等近百家获奖企业以荣获红星奖作为商业广告宣传卖点，赢得了市场份额。

经过严格评审，2009年共有102家企业的191件产品获奖，其中至尊金奖1项，金奖8项，最具创意奖10项，红星奖172项及1个最佳团队和1名最佳新人。另外，为表彰在国庆60周年群众游行中的优秀设计，中国创新设计红星奖委员会特增设“红星奖国庆60周年特别奖”，颁发给优秀国庆彩车设计作品。

2009年是新中国六十华诞，每一个中国人都能感受到社会经济飞速发展的脉搏。当物质生活相对丰富，人们开始追求更有品质的“设计生活”，设计消费市场充满无限活力。面对新时期的机遇和挑战，红星奖任重而道远。

年轻的红星奖刚刚迈出万里长征的第一步，我们自信红星奖的未来一片璀璨，她将与共和国共同成长繁荣，成为影响世界的中国设计巨奖。

红星照耀下的中国，必将是创新的中国。

中国创新设计红星奖委员会
2009年11月16日

Toward New Opportunity of Industrial Design

2009 is the fourth year since the establishment of China Red Star Design Award and also a year in which the Award is progressing. With a jury of international experts, strict and impartial judging procedures, broad engagement range of enterprises, candidate products of high quality, magnificent award ceremony, worldwide promotion and mutual certification with Icsid, the Red Star Award is recognized as an international brand with high credibility.

In 2009, the Red Star Award totally collected 3821 products from 758 enterprises, extending its coverage to 74% of all the Chinese administrative regions. The numbers of enterprises and products increase by 34% and 21% respectively compared with last year. Therein, the category of candidate furniture increases by 146% than last year and the category of equipment by 100%. Companies running for the competition take 76% of the total candidate enterprises, including the consecutive participants like Lenovo, Haier, Vatti, Oulin, Midea, Qumei and Toread, and the design companies which have received Reddot Award like Leo Design, UI Design, Penpower Technology and Yoropen.

Over the past four years, 68 experts from 12 countries and regions like America, Britain, Germany, France, Japan, Korea, Australia, Chinese Taiwan and Hong Kong regions have been the jury members. They are from various fields such as design, science and technology, economy, media and marketing. The international and trans–disciplinary characteristics of the jury ensure the authoritativeness and fairness of the Red Star Award.

The Red Star Award focuses on promoting the brand value, being invited to 18 cities such as Hong Kong, Tianjin, Shanghai, Fuzhou, Xi'an, Nanjing, Qingdao and so on to carry out 28 tour exhibitions. We organized more than 20 interactivities between both domestic and overseas enterprises and local manufacturing factories. We organize promotion and marketing activities for awarded enterprises in tour exhibitions to increase their order numbers.

With the influence of the Red Star Award extending, its social response comes conspicuous. Almost one hundred awarded enterprises like Changhong, Skyworth, Konka and Aigo publicize their products with their being awarded by the Red Star Award as commercials, gaining more market proportion.

Through strict judging, a total of 191 products from 102 enterprises are awarded.In addition, in order to award the excellent designs in the public parade for the 60th anniversary of the new China, We specially adds a Special Prize for the 60th Anniversary for the excellent float design.

In 2009 which marks the 60th anniversary of the new China, every Chinese can touch the pulse of the rapid development of society and economy. When life comes relatively affluent, we begin to pursue "design life" of higher quality, thus design consumption becomes more dynamic. Confronting new opportunities and challenges, the Red Star Award has a long way to go.

The young Red Star Award just takes the first step of a long march. We are confident that the Red Star Award has a bright future; it will flourish in synchrony with China and grows to be an award with worldwide influence.

Red Star over China is innovative China.

China Red Star Design Award Committee

16th November 2009

中国创新设计红星奖组织机构

主办单位
中国创新设计红星奖委员会

发起单位
中国工业设计协会
北京工业设计促进中心
国务院发展研究中心《新经济导刊》杂志社

项目支持
北京市科学技术委员会

承办单位
北京工业设计促进中心

协办单位
中国产学研合作促进会
中国光华科技基金会
北京工业设计促进会
天津市工业设计协会
上海工业设计协会
重庆工业设计协会
黑龙江省工业设计协会
黑龙江省艺术设计协会
江苏省工业设计学会
山东省企业联合会
河南省工业设计协会
湖南省工业设计协会
广东省工业设计协会
四川省工业设计协会
陕西省工业设计协会
无锡工业设计协会
宁波市工业设计联合会
福州市科学技术局
顺德工业设计协会
青岛市工业设计协会
深圳市设计联合会
广州工业设计促进会
珠海市工业设计协会

专业支持媒体
Design News China
Frame China
造车网

协办媒体
新华网传媒频道
北京电视台魅力科学
科技日报
搜狐数码频道
科学时报
北京商报
家电科技
视觉中国
视觉同盟

Institutes of China Red Star Design Award

Organizer
China Red Star Design Award Committee

Sponsor
China Industrial Design Association
Beijing Industrial Design Center
New Economy Weekly

Support
Beijing Municipal Science & Technology Commission

Host
Beijing Industrial Design Center

Co-organizer
China Industry–University–Research Institute Collaboration Association
China Guanghua Science and Technology Foundation
Beijing Industrial Design Promotion Organization
Tianjin Industrial Design Association
Shanghai Industrial Design Association
Chongqing Industrial Design Association
Heilongjiang Industrial Design Association
HeiLongJiang Province Artistic Design Association
Jiangsu Industrial Design Association
Shandong Enterprise Confederation
Henan Industrial Design Association
Hunan Industrial Design Association
Guangdong Industrial Design Association
Sichuan Industrial Design Association
Shaanxi Industrial Design Association
Wuxi Industrial Design Association
Ningbo Industrial Design Union
Fuzhou Municipal Bureau of Science and Technology
Shunde Industrial Design Association
Qingdao Industrial Design Association
Shenzhen Industrial Design Association
Guangzhou Industrial Design Association
Zhuhai Industrial Design Association

Supporting Media
Design News China
Frame China
Zaoche168.com

Cooperating Media
Xinhuanet.com.Media
Beijing Television.Funny Science
Science&Technology Daily
SOHU.com.Digital
Science Times
Beijing Business Today
Household Appliance Technology
ChinaVisual.com
VisionUnion.com

中国创新设计红星奖委员会

主席
朱焘

联席主席
闫傲霜
陈冬
石定寰

执行主席
黄武秀
陈冬亮

委员
刘晖　王海宁　白云祥　刘亚杰
张伟　张锡　张震甫　何人可
宋慰祖　林勇　周砚　郑佰森
郑寿平　郑德湘　胡启志　徐和平
蒋国栋　韩选华　程建新　虞汉悦
韩小军　喻湘晖　樊超然

中国创新设计红星奖办公室

主任
黄武秀（兼）

副主任
宋慰祖（常务）
楼晓红

工作人员
王丽娜　张苗苗　高洁　孔倩倩
许侃　栾一丞　刘爽　侯爱莉

China Red Star Design Award Committee

Chairman
Zhu Tao

Co-Chairman
Yan Aoshuang
Chen Dong
Shi Dinghuan

Executive Chairman
Huang Wuxiu
Chen Dongliang

Members
Liu Hui / Wang Haining / Bai Yunxiang / Liu Yajie
Zhang Wei / Zhang Xi / Zhang Zhenfu / He Renke
Song Weizu / Lin Yong / Zhou Yan / Zheng Baisen
Zheng Shouping / Zheng Dexiang / Hu Qizhi / Xu Heping
Jiang Guodong / Han Xuanhua / Cheng Jianxin / Yu Hanyue
Han Xiaojun / Yu Xianghui / Fan Chaoran

China Red Star Design Award Office

Director
Huang Wuxiu

Deputy director
Song Weizu
Lou Xiaohong

Staff
Wang Lina / Zhang Miaomiao / Gao Jie / Kong Qianqian
Xu kan / Luan Yicheng / Liu Shuang / Hou Aili

2009中国创新设计红星奖评委合影

Group photo of jury of 2009 China Red Star Design Award

中国创新设计红星奖专家委员会

Jury Committee of China Red Star Design Award

柳冠中（中国）

清华大学美术学院责任教授、博士生导师
享受政府津贴学者
清华大学美术学院工业设计系系统设计工作室总设计师
中国工业设计协会副理事长兼学术和交流委员会主任

Liu Guanzhong (China)

Professor & Doctor Tutor of Art College of Tsinghua University
Chief Designer of System Design Studio of Industrial Design Department of Art College of Tsinghua University
Deputy Chairperson of China Industrial Design Association & Director of Knowledge and Communication Committee of China Industrial Design Association

何人可（中国）

湖南大学设计艺术学院院长、教授
中国工业设计协会副理事长
教育部高等学校工业设计专业教学指导分委会主任委员
中国机械工业教育协会工业设计学科教学委员会主任委员

He Renke (China)

Dean & Professor of Design School of Hunan University
Deputy Chairperson of China Industrial Design Association
Chairman of China National Instructive Committee of Industrial Design Education
Chairman of Instructive Committee of Design of China Engineering Education Association

Peter Zec（德国）

2005~2007国际工业设计协会联合会主席
红点设计奖主席
北威设计中心主席

Peter Zec (Germany)

President of 2005~2007 Executive Board of International Council of Societies of Industrial Design (ICSID)
President of Red Dot Design Award
Chairman of Design Zentrum Nordrhein Westfalen

Ralph Wiegmann（德国）

德国IF设计奖常务理事
欧洲设计管理奖评委会主席

Ralph Wiegmann (Germany)

Managing Director of International Forum Design
European Design Management Award Jury Chairman

Kristina Goodrich（美国）

美国工业设计师协会（IDSA）前CEO

Kristina Goodrich（USA）

Former Executive Director & Chief Executive Officer of Industrial Designers Society of America (IDSA)

喜多俊之（日本）

日本G–mark奖评委会前主席

Toshiyuki Kita（Japan）

Former Chairperson of the Jury of G–mark Award

George Teodorescu（德国）

全球联合学院执行理事

TESIGN国际设计咨询公司主席

George Teodorescu（Germany）

Member of Executive Board of United Global Academy

President of TESIGN– International Design Consulting

Chelsea Sutula（美国）

美国工业设计师协会（IDSA）前奖项主任

Chelsea Sutula（USA）

Former Director of Awards & Special Projects of Industrial Designers Society of America (IDSA)

Alberto Cannetta（意大利）

意大利设计协会首席亚洲顾问

原意大利驻华使馆参赞

意大利设计巡展（I.DoT）和意大利经典设计展示作品评委

Alberto Cannetta（Italy）

Chief Asian consultant of Italian Design Agency

Former culture counselor of Italian Embassy in China

Jury of I.DoT and ID_CS exhibition

李一奎 (韩国)

韩国设计振兴院前院长
曾历任中小企业厅技术支援局、创业风险局局长，京畿地方中小企业厅厅长等

LEE Il-kyoo (Korea)

Former President and Chief Executive Officer of Korea Institute of Design Promotion
Serving at Ministry of Commerce, Industry and Energy (MOCIE) for improvement and growth of Korean design industry

Michael Thomson (英国)

伦敦Design Connect公司创始人兼负责人
欧洲设计协会（BEDA）办公署副部长
2001~2005国际工业设计协会联合会执委会委员

Michael Thomson (UK)

Founder and Principal of Design Connect of London
Vice President of the Bureau of European Design Associations (BEDA)
Member of the Executive Board of the International Council of Societies of Industrial Design (ICSID) (2001~2005)

Brandon Gien (澳大利亚)

澳大利亚国际设计奖总裁
澳大利亚标准组织设计与联络部主任

Brandon Gien (Australia)

Director of the Australian International Design Awards and the Design & Communications Division at Standards Australia

赵英吉 (韩国)

韩国Design mall设计公司主席
韩国设计公司协会前会长
韩国设计师协会理事会主任（1999~2000年）
韩国好设计奖评委（2002/2004年）

Cho Young Kil (Korea)

President of Design mall co., Ltd.
Former Chairman of Korea Design Firms Association
Director of Board of Korea Association of Industrial Designers(1999~2000)
Jury of Good Design Award of Korea(2002~2004)

Ola Lantz (瑞典)

CBD国际设计公司首席设计师

Ola Lantz (Sweden)

Chief designer of CBD International Design Company

林衍堂（中国香港）

香港理工大学设计学院副学院主任
英国特许设计学会资深会员
欧洲设计局会员
香港设计局董事

LAM,Yanta（China Hong Kong）

Professor of Design & Associate Head, school of Design, the Hong Kong Polytechnic University
Fellow, Chartered Society of Designers, UK
Member, Bureau of European Designers Association
Director, Hong Kong Design Centre

陈文龙（中国台湾）

2007~2009国际工业设计协会联合会执委会委员
浩汉产品设计股份有限公司总经理

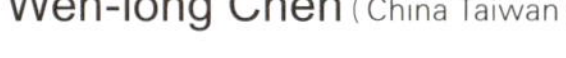

Wen-long Chen（China Taiwan）

Member of 2007~2009 Executive Board of International Council of Societies of Industrial Design (ICSID)
President of Nova Design Co., Ltd.

官政能（中国台湾）

台湾实践大学副校长、产品与建筑设计研究所所长
台湾实践大学设计学院首任院长

Cheng Neng Kuan（China Taiwan）

Deputy Principal of Shih Chien University
Chairman of Institute of Industrial & Architecture Design of Shih Chien University
The First Dean of School of Design of Shih Chien University

陈冬亮（中国）

北京工业设计促进中心主任
北京工业设计促进会理事长
中国创新设计红星奖委员会执行主席
中国工业设计协会常务理事兼专业咨询委员会秘书长
意大利设计中国巡展华人评委
韩国好设计奖评委
清华大学美术学院工业设计发展顾问

Chen Dongliang（China）

Director of Beijing Industrial Design Center
Director of Beijing Industrial Design Promotion Organization
Executive Chairman of China Red Star Design Award Committee
Executive Director of China Industrial Design Association
Jury of Italian Design on Tour
Jury of Korean Good Design Award
Industrial Design Development Consultant of Academy of Arts & Design, Tsinghua University

鲁晓波（中国）

清华大学美术学院教授、博士生导师
清华大学信息艺术设计系主任
中国工业设计协会常务理事
红点设计奖评委

Lu Xiaobo（China）

Professor & Doctoral Tutor of Academy of Arts & Design, Tsinghua University
Director of Information Art & Design of Academy of Arts & Design, Tsinghua University
Committeeman of China Industrial Design Association
Jury of Red Dot Design Award

童慧明（中国）

广州美术学院设计学院教授、院长
中国美术家协会工业设计艺术委员会委员
中国工业设计协会常务理事
广东省工业设计协会副会长

Tong Huiming (China)

Dean & Professor of College of Design of Guangzhou Academy of Fine Arts
Committeeman of Industrial Design Committee of Association of Chinese Artists
Executive Director of China Industrial Design Association
Deputy Chairman of Guangdong Industrial Design Association

朱钟炎（中国）

同济大学建筑与城市规划学院艺术设计系教授、博士生导师
燕山大学、中国美术学院上海设计学院等六所大专院校兼职教授

Zhu Zhongyan (China)

Professor & Doctoral Tutor of Art Design Department of Architecture & Urban Planning Institute of Tong Ji University
Adjunct Professor of Six Chinese Universities including Yan Shan University etc

林家阳（中国）

同济大学设计艺术研究中心主任
教育部高等学校高职高专艺术设计类教学指导委员会主任
中国艺术研究院研究员兼视觉艺术研究委员会主任

Lin Jiayang (China)

Director of Research Center of Design & Art of Tongji University
Chairman of China National Instructive Committee of Art Design Education
Director of the Visual Arts Research Committee of China Art Research Institute

许　平（中国）

中央美术学院设计学院副院长、教授、博士生导师
中央美术学院设计文化与政策研究所所长
中国工业设计协会常务理事

Xu Ping (China)

Deputy Dean & Professor & Doctor Tutor of Design School of China Central Academy of Fine Arts
Director of Design Culture and Strategy Research Center of China Central Academy of Fine Arts
Executive Director of China Industrial Design Association

Martin Darbyshire（英国）

2007~2009年国际工业设计协会联合会执委会委员
英国Tangerine设计公司创始人

Martin Darbyshire (UK)

Member of 2007~2009 Executive Board of International Council of Societies of Industrial Design (ICSID)
Chief Executive Officer of Tangerine Design Co.

Francois Lenfant (法国)

通用电气医疗全球设计经理

Francois Lenfant (France)

Global Design Manager for Europe & Middle-East (EMEA),GE Healthcare

Lee Soon In (韩国)

首尔设计中心主席
韩国弘益大学教授
LG欧洲设计中心执行主任 (1990~1995年)

Lee Soon In (Korea)

President of Seoul Design Center
Professor of Hong-ik University
Managing Director of LG Europe Design Center (1990~1995)

袁　岳 (中国)

零点研究咨询集团董事长
中国信息协会市场研究业分会副会长
北京科技咨询业协会理事长

Victor Yuan (China)

Chairman of the board, Founder and President of Horizon Research Consultancy Group
Vice President of China Marketing Research Association
President of Beijing Consulting Association (BCA)

程建新 (中国)

华东理工大学艺术设计与传媒学院院长、教授、学术委员会主任

Cheng Jianxin (China)

Dean, the School of Art, Design and Media, East China University of Science and Technology

严　扬 (中国)

清华大学美术学院工业设计系教授

Yan Yang (China)

Professor of Industrial Design Department of Academy of Arts & Design, Tsinghua University

邱丰顺（中国台湾）

北京艺有道工业设计有限公司创办人、总经理
摩托罗拉亚洲区设计总监（2001～2006年）

Kumo Chiu (China Taiwan)

President of Idea Dao Design Ltd.
Asia Design Director of Consumer Experience Design, Mobile Device, Motorola(2001~2006)

Max Burton（美国）

青蛙设计公司执行创意总监
前耐克公司技术实验室创意总监

Max Burton (USA)

Executive Creative Director, Frog Design
Former Creative Director for Nike's Tech Lab

David Kester（英国）

英国设计委员会主席（2003年至今）
英国D&AD主席（1994～2003年）
英国皇家艺术学院执委
设计商业协会特别顾问

David Kester (UK)

Chief Executive of Design Council (since 2003)
Chief Executive of D&AD (1994~2003)
Council member of the Royal College of Art
Special advisor to the Design Business Association

田中一雄（日本）

GK设计株式会社总裁
2007～2009国际工业设计协会联合会执委会委员

Kazuo Tanaka (Japan)

President of GK Design Group Inc.
Member of 2007~2009 Executive Board of International Council of Societies of Industrial Design (ICSID)

Park Youngsoon（韩国）

韩国设计协会联盟主席
韩国设计振兴院理事
韩国好设计奖评委会主席（2005年）

Park Youngsoon (Korea)

Chairman of Korea Federation of Design Associations (KFDA)
Director of Korea Institute of Design Promotion (KIDP)
Chairman of Jury committee in Korea Good Design Award (2005)

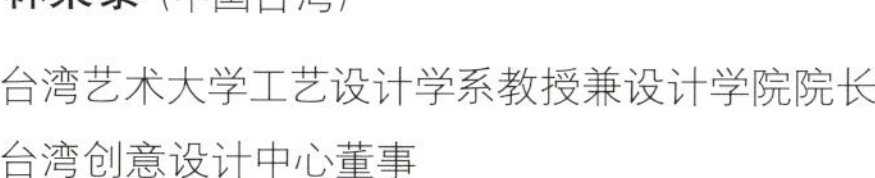

林荣泰（中国台湾）

台湾艺术大学工艺设计学系教授兼设计学院院长
台湾创意设计中心董事

Rungtai Lin (China Taiwan)

Professor, Department of Crafts and Design, Taiwan University of Arts
Dean of Design College, Taiwan University of Arts
Director Board of Taiwan Design Center

胡　越（中国）

全国建筑设计大师
北京市建筑设计研究院总建筑师
中国建筑学会理事

Hu Yue (China)

National Architectural Design Master
Chief Architect of Beijing Institute of Architectural Design
Director of Architectural Society of China

殷正声（中国）

同济大学设计创意学院教授
中国工业设计协会常务理事
上海工业设计协会副理事长

Yin Zhengsheng (China)

Professor of Design Innovation College of Tongji University
Executive Member of the Council of China Industrial Design Association
Vice President of Shanghai Industrial Design Association

刘亚东（中国）

科技日报副总编辑
曾任科技日报驻联合国暨纽约首席记者5年
2008年获第九届长江韬奋奖（长江系列）

Liu Yadong (China)

Vice Editor of Science and Technology Daily
Chief Journalist of Science and Technology Daily in UN New York
Granted Changjiang Taofen Award in 2008

至尊金奖

Best of the Gold Prize

中国创新设计红星奖

China Red Star Design Award

09

中国电信股份有限公司上海研究院

Shanghai Research Institute of China Telecom Corporation Limited

九宫盲人手机

让残障人士平等地享受现代通信技术带来的便利，与社会建立起人性化的沟通平台是“九宫盲人手机”设计的出发点。通过九宫格界面，将整个手机界面的交互操作规定在3×3的九个点上，同时利用中心点和导向轨道引导盲人的手指操作，方便了盲人对于界面的认知和操作；盲人视频功能可以让健全人通过自己的手机屏幕看到盲人手机所拍摄的图像，利用中国电信的移动全球眼平台，通过手机视频通话功能，实现对盲人的远程帮助。本产品的设计充分考虑到了盲人的使用情况，比如：不同触感材质的功能区域划分，利用滑盖形式进行上锁、解锁键盘的操作，方便使用的充电和数据接口，方便的佩戴方式等。本产品携带方便，操作简单，充分体现了设计师对盲人的关爱和社会责任感。

9 Spots “Mobile Global Eye”

This product gives the blind an opportunity to take advantage of the convenience of modern communication technologies. It aims to establish a humanistic communication platform between the blind and society. With the nine–square grid interface, the operation is restricted to nine points. Meantime, it uses a central point and tracks to guide the fingers of the blind, facilitating recognition and operation of the blind. The video function enables the others to see the pictures shot by the product. The distant assistance to the blind is realized resorting to the platform of "Motion Global Eye" developed by China Telecom and the visual conversation. This product is handy with a sole operative, manifesting designer's care and social responsibility to the blind.

speaker.
盲人可直接对准门口说话.
机器会自动录音.
凸起可以,
帮助盲人
确定机器的
正反.
139176
1301

MNO
JKL
WXYZ
GHI
TUV
PQRS
09.06.09 7:00 am
09.06.09 1:30 pm
09.06.09 6:08 pm
09.06.09 1:03 pm
09.06.09 4:16 pm
09.06.09 3:30 pm

金奖

Gold Prize

中国创新设计红星奖

China Red Star Design Award

09

康佳集团生活电器

空气净化时尚型

本产品强调对空气感应的敏锐度，以融入环境的方式，缓和但明显地提升空气质量，同时营造出清新的环境氛围；将面板操作减至最直接的状态，让用户不需要在众多按键中选择；开阖与更换动作都隐藏在侧面，更换滤网只要将侧板卸下，简单易用。其设计理念是要传达一种透明、清澈的感觉，在设计上尽量将线条精简到最协调的状态。材质上使用两种材质互相搭配，以纯净白色作为机底的整体配色，搭配雾面的铝、绝对的黑，简约的生活态度在此展现。

KONKA GROUP Co., Ltd.

Air Cleaner

This product emphasizes its acute response to air, markedly promotes air quality and builds clean environment. It decreases interface operation to a most direct way, facilitating the user's choice among many keys. Designer aims to convey a sense of transparency and clearness. Two kinds of materials are coordinated with pure white as its main tone and black as supporting color. Thereby, it shows a simple life style.

KONKA

太仓市贤信堂设计咨询有限公司

SINCERE TONE DESIGN AND CONSULTING CO., LTD

ORIGIN-多功能城市防灾应急微型勤务车

Mini Multifunction Emergency Vehicle

本产品主要是遇到自然灾害或发生其他重大事件时使用的应急机动设备，可扩展的微型通用移动平台，将短距离交通、消防、临时物流、安防巡逻、小马力牵引、破冰除雪、路面清扫等功能模块方便高效地整合起来，为社区、居民点、临时营区、楼宇大型室内环境、会展体育馆等人流集中而应急设备移动不便的局部区域提供临时的应急服务。它便于携带，操作简单，耐久力强，既实用又美观的特性一定会使它的驾驶者爱不释手。

This product is designed to meet an emergency like natural disaster or other disastrous events. It effectively coordinates the functional modules of short-distance transport, fire fighting, temporary distribution, security and patrolling, petty horsepower traction, breaking ice and cleaning snow and ground, etc. It provides temporary expedience for crowded local places where it's inconvenient to move emergent facilities like social communities, resident areas, temporary battalions, large-sized interior environment, exhibition centers and stadiums. It's handy with a strong hardiness and simple ways of operation. The practicality and aesthetics will endear this product to its drivers.

ORIGIN

中兴通讯股份有限公司

“睿智” —— ZTE中兴T8000高端路由器

T8000设计大胆，体现出设计者高超的设计能力。它整体造型上利用大块面的转折，将普通插箱上的零碎单元有机地组织在一起，形成一个整体，并因此呈现出山岩般的阳刚气质和干净利落的效果，体现出专业感。块面之中利用条状线型配以网板，使设备在大气之余不失细腻和精致，具有品质感。整体以银黑搭配，辅之以蓝色灯光，设计美观、简单大方，合理和进步的美学设计使T8000具有实用的功能性。

ZTE CORPORATION

“Sagacity” —— ZTE T8000 High-end Router

The T8000 is a bold design that reflects the awesome power and intelligence it houses. It integrates the fragmentary components of a common subrack into an organic whole, showing masculinity, cleanness and specialty. With grey black as its main tone and blue as assistance, the aesthetic treatments are appropriate and yet progressive, making the T8000 functional.

ZTE中兴

上海木马工业产品有限公司

Shanghai Moma Industrial Product Design Co., Ltd.

飞利浦–我的阅读灯

My Reading Light

该产品设计理念先进，LED技术让我们随时随地享受阅读的乐趣，尤其还可以在飞机上使用。只照亮书页又不打扰周围人群的功能，可以保护每个人的私有空间，很人性化。该产品操作简单，形似书签，实用性强，开创了一种新的读书方式。

This product is sophisticatedly designed. The LED technology gives us an opportunity to enjoy our reading, especially it can be used on an airplane. It humanistically protects privacy of everyone by merely illuminating book page and without disturbing the surrounding people. It's a handy and practical product, shaped like a bookmarker, which starts a new reading way.

PHILIPS

北京蓝威科瑞科技有限公司

火鹰红外热成像仪

该产品是应用于消防、核振领域的红外热成像设备。它体积轻巧，易于操作，黑色手柄的曲面造型便于手持。整个产品使人感觉坚固耐用，除手持外还配有三角架能使产品稳定使用。它利用红外成像的特点，可以轻易地在烟雾、黑暗的环境下探测到处于危险中的人员，从而实施救援；它还可以检测到有安全隐患的发热设备，在发生危险之前采取措施。考虑到设备工作的恶劣环境，"火鹰"具有较高的防护设计，外层附有包胶，防止碰撞、意外跌落时对设备造成的损坏。

Beijing Lever Create Co., Ltd.

Firehawk

This product is intended to be used in fire fighting and nuclear vibration as an infrared imaging equipment. The curved shape of its black handlebar makes it easy to handle. It can easily detect lives in danger in smoggy or dark situations. It can also detect the hidden dangers of heating machine so that the dangers can be eradicated before they really happen. Considering the rotten environment under which the product is intended to use, Firehawk has its special consideration in protection design. It is covered with gel to avoid the damage caused by collision and accidental fall.

XHZ UNI·IM
INFRARED SYSTEM
XHZ UNI·IM

北京曲美家具集团有限公司

BEIJING QUMEI FURNITURE GROUP CORP., LTD.

耳朵沙发

Ear Sofa

本产品在普通沙发的基础上增加了三个耳朵，既具创意性又有独特性。耳朵模样的垫子下半部分宽，越往上越薄的设计形式，增加了使用者的安全感，并且不需要另外特别的装置就可以固定垫子的位置。根据垫子摆放的位置不同，可以让更多的人坐到沙发上，并且摆出更多不同的姿势，更好地满足使用者的需求：晚上可作为床使用，还可放在儿童房内作为家居通用产品。黑色底座配上金黄色的垫子更加突出了沙发的优雅。本品上市后深受儿童、时尚人士的青睐，获得了广泛的市场好评。

It innovatively and uniquely adds three ears on the basis of a common sofa. The ear-shaped cushion enhances safety sense of its users with its broad lower part and increasingly thin to its top. According to different positions of the three cushions, more people can sit on it with different positions. It can be used as bed at night and sofa in living room. Its black base coordinated with golden cushions shows more elegance. It gets great popularity among children and fashionable people after its going to the market.

深圳市浪尖工业产品造型设计有限公司

ONYX电子阅读器

ONYX电子阅读器的设计哲学体现了中国传统文化中相互关爱、以人为本的精神和中国传统"天圆地方"的思想。动为阳、静为阴，故而"天圆"就成了阳的象征，也是本产品灵感和造型的来源。圆形的双面操作界面，把复杂繁琐的按键集中在一起，让操作更简单快捷，带来全新的操作体验。竖屏横屏显示实现了人们左手或右手单独操作的愿望，传达了一种关爱特殊人群的精神。产品巧妙地运用了透明材料的设计，正如中国"滴水石穿"成语中所传达的寓意：读书学习要有恒心和耐心。

ONYX电子阅读器把众多书籍收集其中，具有笔触滑动翻页、互联网浏览、全屏触摸手写等功能；9.5mm的超薄机身，非常方便携带。

SHENZHEN ARTOP Industrial Design Co., Ltd.

ONYX BOOX

The design of ONYX Electronic Reader embodies the ideas in the traditional Chinese culture such as mutual love, humanity and "Round heaven and square earth". Motion stands for Yang and motionlessness stands for Yin, so "round heaven" stands for Yang which also inspires the design of this product. The round double operation interface collects the complex keys together, facilitating the operation of this product. Its vertical and horizontal screen helps to realize people's wish to operate with right hand or left hand alone, conveying its care to special people. This product is composed of transparent materials, echoing the moral of the Chinese idiom "Constant dripping wears away a stone": study requires perseverance and persistence.

I Have a Dream
by Martin Luther King, Jr.
I am happy to join with you today in what will go down in history as the greatest demonstration for freedom in the history of our nation.
Five score years ago, a great American, in whose symbolic shadow we stand today, signed the Emancipation Proclamation. This momentous decree came as a great beacon light of hope to millions of Negro slaves who had been scared in the flames of withering injustice. It came as a joyous daybreak to end the long night of their captivity.
But one hundred years later, the Negro still is not free. One hundred years later, the life of the Negro is still sadly crippled by the manacles of segregation and the chains of discrimination. One hundred years later, the Negro lives on a lonely island of poverty in the midst of a vast ocean of material prosperity. One hundred years later, the Negro is still languished in the corners of American society and finds himself an exile in his own land. And so we've come here today to dramatize a shameful condition.
ONYX
MENU
NEXT
BACK
PREV
OK

北京品物堂产品设计有限公司

PER Design

专业耳麦PT301

该产品造型时尚，材料及色彩搭配得当，充分体现了专业耳机的个性，设计中亮点突出，折叠骨架设计的包装盒携带方便。它将新鲜概念一气呵成地表现出来，赋予自身强大的专业面貌及鲜明的性格特征，硬朗的边角及柔和的直面过渡，耿直中透露着低调的含蓄。另外，该产品的工艺及各活动部位的连接细节恰到好处，给人留下深刻印象。

Professional Earphone

This product is fashionable with appropriate coordination of materials and colors. Its folding package box makes it portable. Its straight angles and soft transitions make it appear honest and frank but a little low-pitched implicity. It gives its observers a deep impression that its components are appropriately connected.

POWER
TELIKOU
000020281

最具创意奖

Best Originality Prize

09

中国创新设计红星奖

China Red Star Design Award

齐思工业设计咨询（上海）有限公司

GSR 迷你专业螺丝电起子

本产品定位明确，为专业人士设计，是世界上已知同类产品中最小的。设计师在握手部分采用了人性化操作的曲线，并运用了软胶材料，握起来感觉很舒服。产品头部安装的LED聚光照明灯，可以在黑暗的环境中轻易地找到施工部位。该产品还有充电锂电池、软包装等辅件，整个产品及备件都小型化，便于携带，充分体现迷你的特点。该产品很好地体现了BOSCH的设计语言，有很强的品牌设计意识。

TEAMS Design Consulting Co., Ltd.

GSR ProDrive Professional

This product is specially designed for professionals and is the smallest among similar products in the world. It feels comfortable with its humanistic curve in the handle and the material of soft gel. The LED spotlight installed in the front of this product makes it easy to find construction place in dark environment. With lithium batteries and soft package, this product and its components are miniaturized to be portable. It embodies the design language of BOSCH well with a strong sense of brand design.

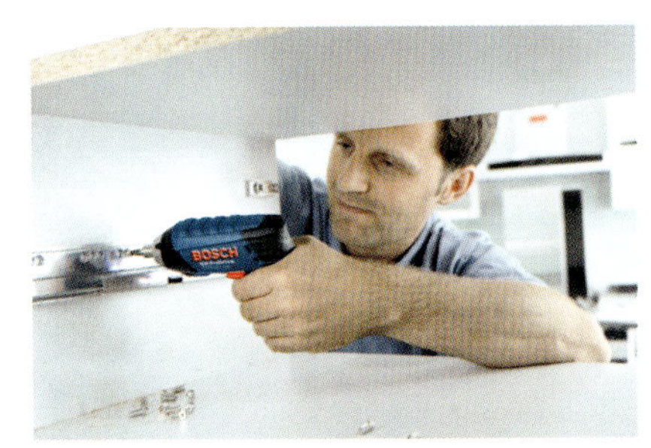

BOSCH
GSR ProDrive
Professional
BOSCH
GSR ProDrive
Professional

桔思创意设计

U.I.Design Co., Ltd.

香料娃娃(油醋瓶)

油醋瓶是人们日常生活中最常见的产品，KeruKeru香料娃娃之所以让人眼前一亮，就是因为它竟然可以被设计得如此富有情调，同时又大大方便了我们的生活。针对油、醋等液体，KeruKeru香料娃娃可以将调味料先倒入开口的勺子中，方便控制酱料酌量增减，轻松决定食物的口味，同时也为厨房增加了高雅的新成员。

KeruKeru

Oil and vinegar bottles are common things in our daily life. The reason why the Spice Dolls are amazing is that they are designed both practically and stylishly. Liquid seasoning like oil and vinegar can be first poured into the spoon to control its quantity, and thus to control the food flavor. At the same time, the Spice Dolls can add some elegance to the kitchen.

60 ml
50
40
30
20
15
10 ml
2 oz
1 oz

金唐立鑫（北京）科技发展有限公司

JTLX BEIJING Co., Ltd.

体重指数尺/心衰尺

这是一种计算人体体重指数和心功能指数的计算尺，是一款真正的创意设计产品。普通人使用它可以迅速简便地对自己的健康状况进行测量，易于操作。这款产品已经获得了国内发明专利，售价仅为35元，物美价廉。计算尺整体设计风格鲜明、大方，表面的色彩和质感给人以协调统一和轻盈明快的感受。

Weight Index Slipstick / Cardiac

This is a really innovative ruler which is intended to measure human weight and heart function. It can be easily operated by common people to measure their health condition. This product has already got a domestic patent with a retail price of RMB 35 yuan. It's a bold design with lively and harmonious color and quality.

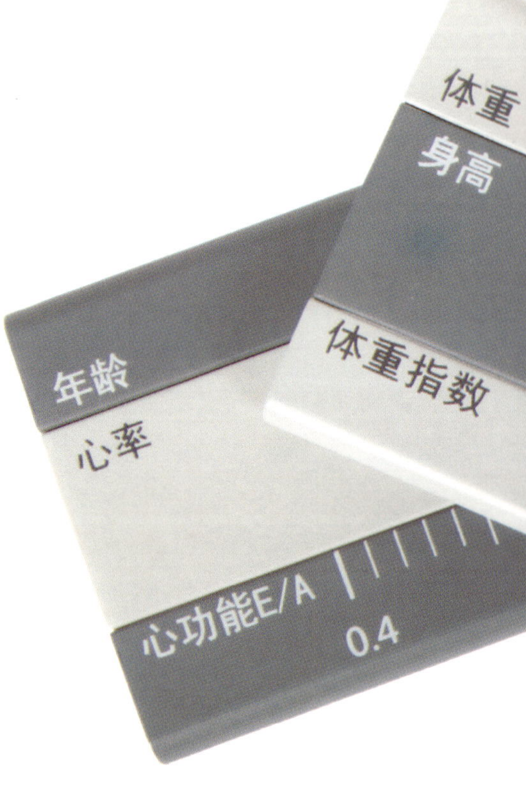

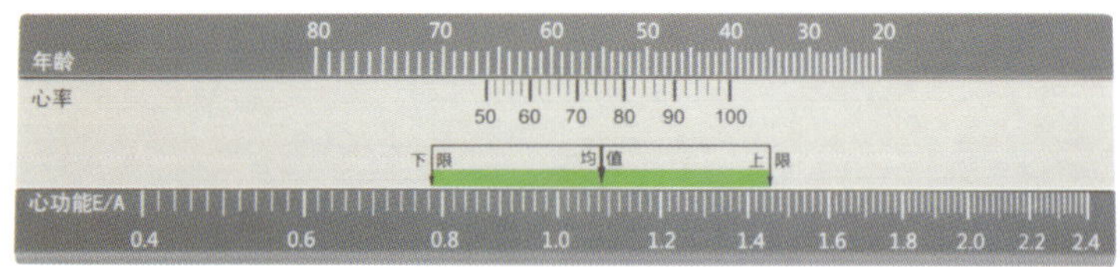

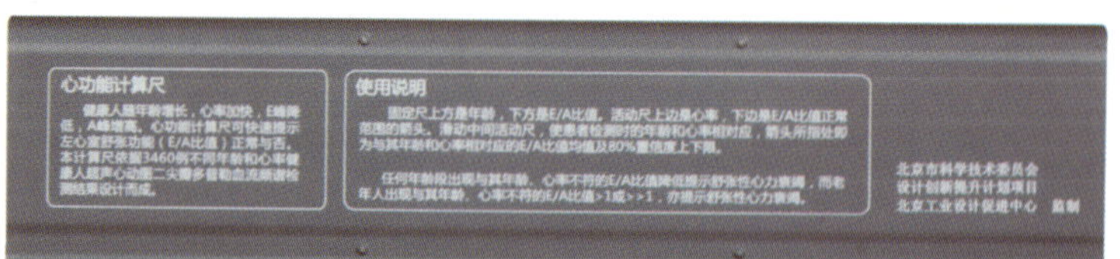

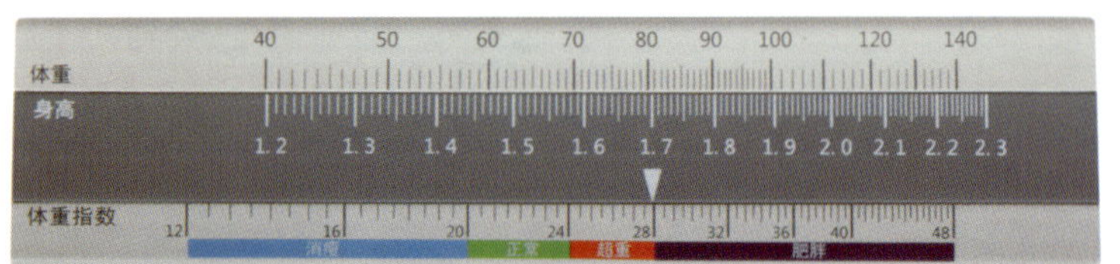

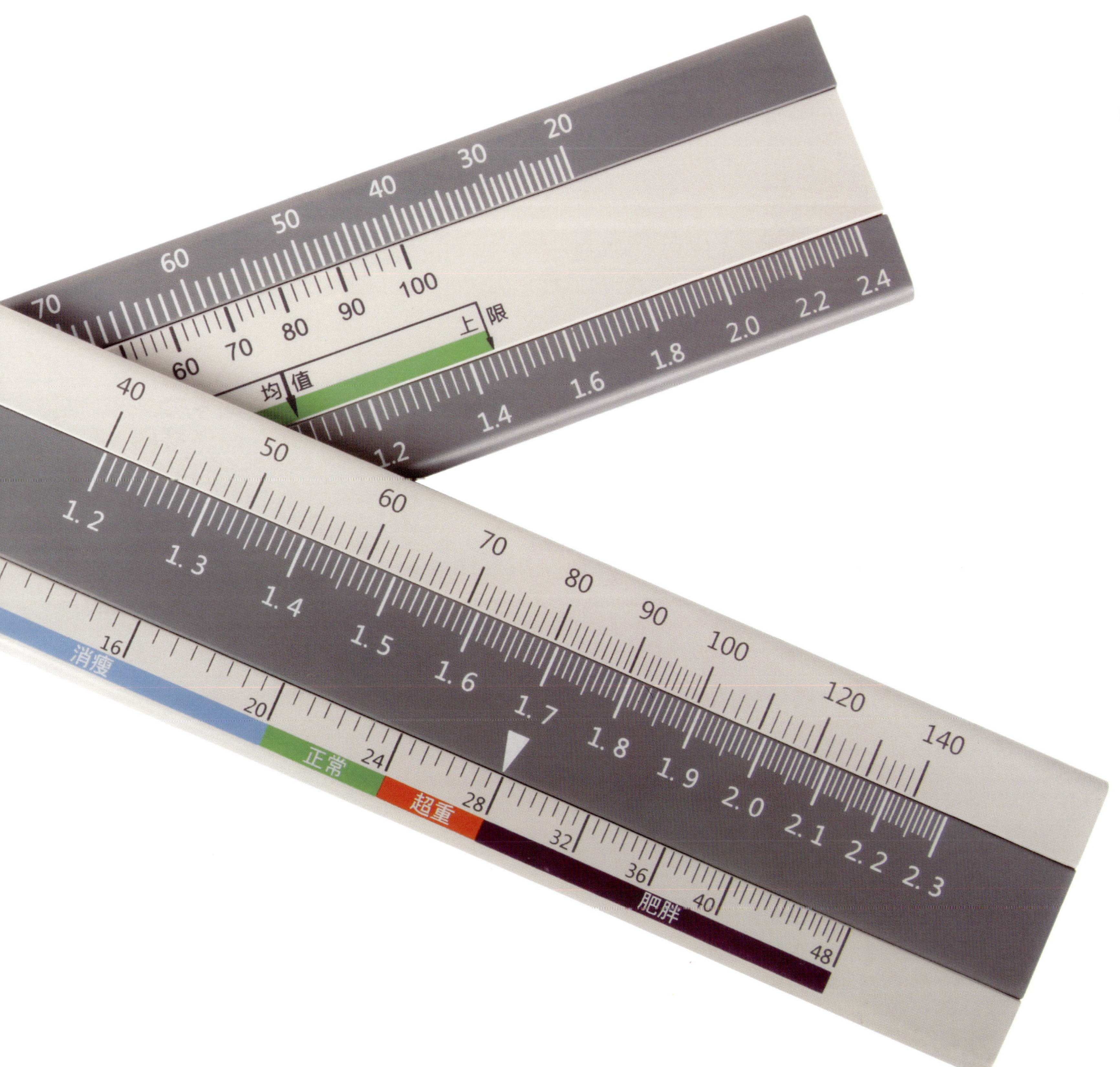

上限
均值
消瘦
正常
超重
肥胖

上海龙域设计

LOE DESIGN

彩色电动剃须刀

USB接口的充电方式，减少了产品的体积与重量，方便携带，特别是对于年轻人和旅游者而言，该产品正迎合了他们的需求。此套电动剃须刀外观新颖，丰富的产品曲线，高光泽的材质与多种时尚轻快的色彩搭配，进一步增强了产品的活力，扩大了消费者的选择面。大刀头的设计，明显提升剃须效果。

Colorful Electric Shaver

The charge way with USB interface decreases the size and weight of the product, making it portable. It especially caters for the youth and the travelers. The liveliness of the product is promoted with its original outward appearance, rich curves, shiny material and fashionable color match, extending the coverage of its customers.

洛可可工业设计公司

LKK Industrial Design Co., Ltd.

汉王电纸书

电纸书正在替代纸介书籍进入广大读者的生活。该设计简洁、精致，全键盘的功能与布局恰到好处，特别是电源键的设计新颖独特，增添了产品的活泼感，让人爱不释手。同时，该产品最大限度地压缩了产品的重量和体积，方便携带，符合主要面向年轻学生的人群定位。

Hanvon E.pbooks

Electronic book is entering our daily life, substituting paper books. This is a simple and exquisite design. Its qwerty keyboard and original power keys add its liveliness and will endear it to its users. It's handy in the sense that its weight and size are maximally condensed. It is consistent with its intended users of young students.

汉王
Hanvon
The Secret Life of Bees
boredom. Thursday afternoons were usually a big peach day, with women getting ready for Sunday cobblers, but not a soul stopped.
T. Ray refused to let me bring books out here and read, and if I smuggled one out, say, *Lost Horizon*, stuck under my shirt, somebody, like Mrs. Watson from the next farm, would see him at church and say, "Saw your girl in the peach stand reading up a storm. You must be proud." And he would half kill me.
What kind of person is against *reading*? I think he believed it would stir up ideas of college, which he thought a waste of money for girls, even if they did, like me, score the highest number a human being can get on their verbal aptitude test. Math aptitude is another thing, but people aren't meant to be overly bright in everything.
I was the only student who didn't groan and carry on when Mrs. Henry assigned us another Shakespeare play. Well actually, I did *pretend* to groan, but inside I was as thrilled as if I'd been crowned Sylvan's Peach Queen.
Up until Mrs. Henry came along, I'd believed
play. Well actually, I did *pretend* to groan, but inside I was as thrilled as if I'd been crowned Sylvan's Peach Queen.
Up until Mrs. Henry came along, I'd believed
PAGE UP
PAGE DOWN
MENU
SYM
DEL

厦门人水卫浴有限公司

Xiamen Renshui Industries Co., Ltd.

禅・静水龙头

Zen・Quiet Faucet

本产品所有操作均采用触摸式控制，具有后工业化社会的特质；具备停水时自动关闭的保护功能，即冷水突然停水，整个供水系统仅有热水，为了避免热水烫伤而直接关闭龙头出水的保护措施，使产品的人机功能更加完美。人在沐浴的过程中，身体和心灵同时得到享受，正可谓"禅则静也"。产品在造型上给人以沉稳和厚实之感，彰显了高贵的品质感。

This product is controlled by touch, carrying the property of post-industrialistic society. It has the protective function of automatic close when water supply is stopped, that is, when cool water supply is suddenly ended, only hot water is supplied, the tap will automatically be closed to avoid scald. Thereby, the ergonomics of the product is well. Its shape looks steady and solid, showing its noble quality.

桔思创意设计

U.I.Design Co., Ltd.

柠檬挤

该产品是专为解决生活中挤柠檬容易沾手的问题而设计的，充分考虑人机工程关系，结构简单，实用性高。采用环保的可降解材料，可再生循环利用。本产品色彩丰富多样，造型美观，增加了用餐的趣味性。

Lemon Run

This product is especially designed to solve the problem that lemon is easy to stick to hand when being squeezed. Thus it is highly ergonomic and practical. It is composed of degradable materials which is recyclable. Its rich color and aesthetic framework add interest into having meals.

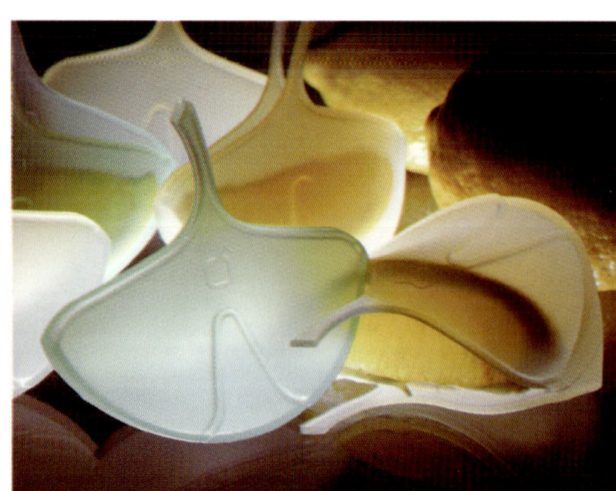

北京博蓝士科技有限公司

模块化可调光度角LED灯具

本产品外形简单大方，是街边一道独特的风景。另外，该产品构造合理，有很多角度，并且可以加长，兼具美观性和实用性；同时通过模块化的设计，使产品使用灵活，制造简单，成本降低。

IDCdesign

"Airflow"—Modular LED Light With Adjustable Beam Angle

This product is a unique street scope with a bold design. It has a proper structure and many perspectives which are both aesthetically and practically designed. Its modular design facilitates its manufacturing and cuts down its cost.

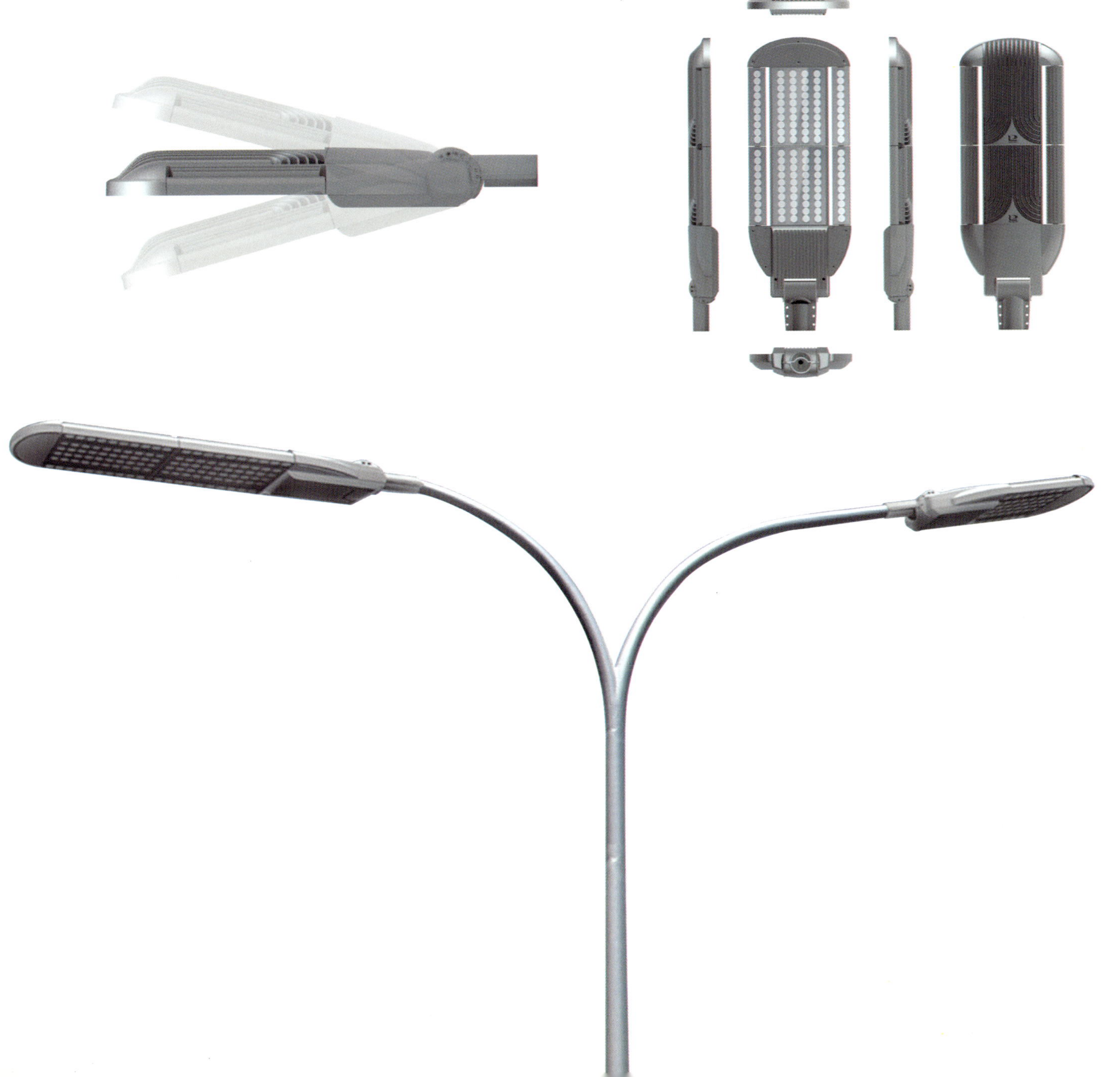

深圳市中兴移动通信有限公司

无线上网卡AC2766

这是一款数据网卡，设计时尚、精巧，具有现代感，是当下生活中的热门产品；本品功能结构合理，采用可回收材料，能再生循环利用，符合环保的要求；材质表现的整体感良好，适合批量生产，符合工艺性要求；产品造型流畅，线条优美，充分体现了时尚的美感。

SHENZHEN ZTE MOBILE TELECOM Co., Ltd.

ZTE-AC2766

This is a digital network card which is fashionably and exquisitely designed. It's properly structured with recyclable materials, so it is in accordance with environmental conservation. It is suitable for mass production and consistent with aesthetic requirement.

ZTE中兴

北京故宫宫苑文化发展有限公司

Beijing Imperial Court Cultural Development Company Ltd.

便携式“皇家普洱”

Royal Pu’er Tea in Square Pieces

皇家普洱茶融入了中国传统文化的内涵，包括宫廷陶瓷、宫廷纹饰、乾隆十骏犬图等图案，增强了其观赏性及收藏价值；它最吸引人的地方就在于其精良的制作工艺和对细节的高关注度；茶块形似巧克力条，直观地告诉人们泡茶的方法；其小铁盒的包装形式，大大提高了产品的便携性和易用性。

The Royal Pu'er Tea is full of elements of the traditional Chinese culture such as royal china and royal patterns, promoting its value of ornament and collection. Its most attractive point consists in its exquisite craftsmanship and its high attention to details. It's formed in the shape of a chocolate bar, visually telling people how to make it. Its package in small iron boxes makes it portable and easy to use.

皇家普洱
ROYAL PU'ER TEA

皇家普洱
ROYAL PU'ER TEA

皇家普洱
ROYAL PU'ER TEA

皇家普洱
ROYAL PU'ER TEA

皇家普洱
ROYAL PU'ER TEA

皇家普洱
ROYAL PU'ER TEA

皇家普洱
ROYAL PU'ER TEA

皇家普洱
ROYAL PU'ER TEA

皇家普洱
ROYAL PU'ER TEA

皇家普洱
ROYAL PU'ER TEA

皇家普洱
ROYAL PU'ER TEA

国庆60周年特别奖

Special Prize for the 60th Anniversary of the Founding of PRC

中国创新设计红星奖

China Red Star Design Award

最佳策划奖

Best Planning Award

星星火炬彩车

星星火炬彩车以立体少先队队徽为主体形象。队徽象征着少先队按照党指引的方向勇往直前，寓意中国特色社会主义事业后继有人。

设计规格：10m×6m×10m

Starry Torch Float

The Starry Torch Float features a three-dimensional Young Pioneers Emblem which symbolizes that the Young Pioneers advance in accordance with the guide of CPC, signifying there are qualified successors for the socialist program with Chinese characteristics.

Design specification: 10m×6m×10m

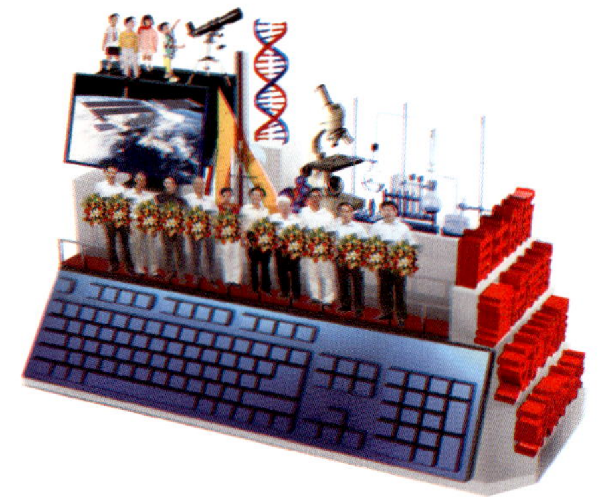

科技创新彩车

科技创新彩车以计算机与实验室工作台为主体形象，彩车前部展示"自主创新，重点跨越，支撑发展，引领未来"的十六字方针，彩车两侧站立着为自主创新领域做出突出贡献的科技工作者代表。

设计规格：15m×7m×10m

Scientific and Technical Innovation Float

The Scientific and Technical Innovation Float features computers and lab workbenches. The guideline of "independent innovation, focus breakthrough, supportive development, guiding the future" is displayed in the front of the Float. In the two sides of the Float stand the representatives of outstanding scientific and technological workers who have made great contribution in the field of independent innovation.

Design specification: 15m×7m×10m

文化成就彩车

文化成就彩车以花瓣底座和中国传统折扇作为主体形象，象征文化事业百花齐放，折扇屏面采用LED屏显示出新闻出版、文物保护、广播电视、非物质文化遗产等领域取得的成就画面。

设计规格：15m×9m×10m

Cultural Development Float

The Cultural Development Float features petal base and folding fans of the traditional Chinese style, standing for the flourishing of the cultural program. The fan Led screen displays the achievements made in the areas such as press, protection of cultural relics, broadcast and TV and intangible cultural heritage.

Design specification: 15m×9m×10m

天津——滨海新貌

滨海新貌彩车以“滨海号”的船造型、海浪图案、祥云60造型等为主体形象，重点展示滨海新区建设成就和辉煌成果，表现天津的城市形象以及飞速发展的整体面貌。

设计规格：15m×6m×10m

Tianjin Float

The Tianjin Float features a model of the boat named "Binhai Hao", pattern of sea waves and the lucky clouds, focusing on displaying the achievements made in the construction of the Binhai New Area, the city image of Tianjin and the rapid development of Tianjin.

Design specification: 15m×6m×10m

山西——魅力山西彩车

魅力山西彩车以太行山为主体形象，夸父追日等神话传说及代表物质文化遗产的五台山等元素，结合经济发展成就塔山煤矿储煤罐的建成，展示山西建设现代文明取得的辉煌成就。

设计规格：15m×6m×10m

Shanxi Float

The Shanxi Float features the Taihang Mountain. Elements like the myth and legendary tale of "Kuafu chasing after the sun" and the Wutai Mountain representing tangible cultural relic, together with the economic achievements and establishment of coal–saving pot in Tashan Coal Mining Company, display the splendid achievements made by Shanxi Province in the construction of modern civilization.

Design specification: 15m×6m×10m

上海——腾飞上海彩车

腾飞上海彩车以一艘驶向未来的船为主体形象，托起市花白玉兰、上海建筑群、世博会吉祥物海宝等元素，预示着上海正朝着建设国际金融中心和国际航运中心的目标快速发展。

设计规格：15m×6m×10m

Shanghai Float

The Shanghai Float features a boat running towards the future, together with elements like the city flower magnolia, Shanghai buildings and the mascot of the World Expo Haibao, foretelling that Shanghai is rapidly developing towards the aim of becoming an international financial center and international shipping center.

Design specification: 15m×6m×10m

浙江——钱江明珠彩车

钱江明珠彩车以水的喜悦、喷泉的欢腾及珍珠的晶莹为主体形象，展现国庆的喜悦和人民的欢庆。

设计规格：15m×6m×10m

Zhejiang Float

The Zhejiang Float features ecstatic water, jubilant fountain spring and crystal pearls, indicating the festivity and joy of the people for celebrating the 60th anniversary of the founding of the People's Republic of China.

Design specification: 15m×6m×10m

重庆——三峡放歌彩车

三峡放歌彩车以三峡大坝和移民浮雕为主体形象，左右两侧各一条标语，均为："三峡百万移民建设新库区"。展现三峡移民开创新生活的精神风貌。

设计规格：15m×6m×10m

Chongqing Float

The Chongqing Float features the Three Gorges Dike and migrant relief, with a same slogan in each side saying "millions of the Three Gorges migrants construct new reservoir area", displaying that the Three Gorges migrants work hard to establish new life.

Design specification: 15m×6m×10m

青海——大美青海彩车

大美青海彩车以高原冰川、至清水源、千里牧场、百里花海为主体形象，表现开放奋进的青海。

设计规格：15m×6m×10m

Qinghai Float

The Qinghai Float features plateau, glacier, pasture and flower sea, expressing an open and dynamic image of Qinghai.

Design specification: 15m×6m×10m

香港——紫荆盛放彩车

紫荆盛放彩车以紫荆花为主体形象，车头部位的金紫荆造型"永远盛开的紫荆花"寓意"一国两制"在香港的成功实践和香港的长期繁荣发展。

设计规格：15m×6m×10m

Hong Kong Float

The Hong Kong Float features Chinese redbud. The golden redbud in the head of the Float——"permanently blooming redbud" signifies the successful practice of "One country, two systems" in Hong Kong and the everlasting prosperity of Hong Kong.

Design specification: 15m×6m×10m

最佳创意奖

Best Originality Award

农业成就彩车

农业成就彩车以充气造型、杂交水稻、温室大棚、新型农业机械、中国黑白花奶牛作为主体形象，表现我国在新型农业科技方面的最新成果。

设计规格：15m×7m×10m

Agricultural Development Float

The Agricultural Development Float features hybrid rice, greenhouse, new agricultural facilities and black and white cow, conveying the latest achievements in new agricultural technology by our country.

Design specification: 15m×7m×10m

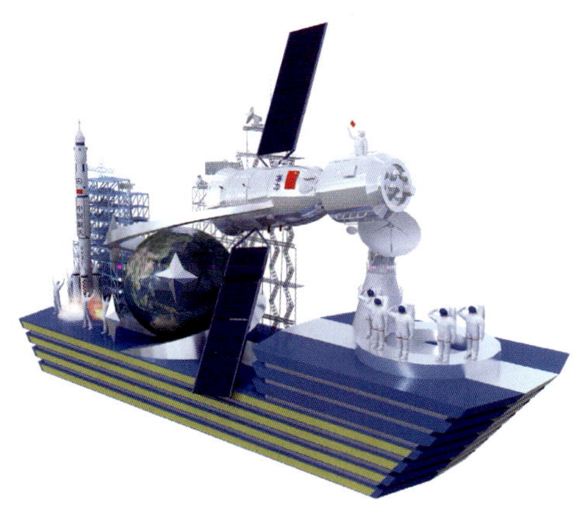

神舟飞天彩车

神舟飞天彩车以火箭、神七飞船、火箭发射架为主体形象，经过天安门时，宇航员升出舱外，伴着《红旗飘飘》的音乐声挥舞五星红旗。

设计规格：25m×7m×15m

Shenzhou Spacecraft Float

The Shenzhou Spacecraft Float features rocket, Shenzhou VII spacecraft and rocket launcher. When the Float passes through the Tian'anmen Square, with the music of "Red Flag Fluttering", the astronaut rises out from the module and waves the Five-starred Red Flag.

Design specification: 25m×7m×15m

我的中国心彩车

我的中国心彩车以大型立体中国结造型为主体形象，底部浪花造型代表着我们的母亲河——长江和黄河，寓意祖国大地对每个中华民族血脉同胞的哺育之情。

设计规格：15m×7m×10m

Patriotic Float

The Patriotic Float features a large-sized Chinese knot. The waves in the base of the Float represent our two mother rivers——the Yangtze River and the Yellow River, denoting the nurture of our motherland to every Chinese compatriot.

Design specification: 15m×7m×10m

北京——北京之歌彩车

北京之歌彩车以天坛、长城、故宫大门、鸟巢、现代建筑群为主体形象，重点展示古老与现代的北京和首都人民正共同谱写一首“人文北京、科技北京、绿色北京”的奋进之歌。

设计规格：15m×6m×10m

Beijing Float

The Beijing Float features the Heaven Temple, the Great Wall, the gate of the Forbidden City, the Bird's Nest and the modern architectural buildings, focusing on displaying the old and new Beijing and that the Capital people are striving for the aim of "People's Beijing, High-tech Beijing, Green Beijing".

Design specification: 15m×6m×10m

江苏——吉祥如意彩车

吉祥如意彩车以如意为主体形象，前部是体现江苏特色的苏州园林，后部由现代建筑、高速公路、电子、化工、机械等建筑模型组合而成，一座现代化斜拉大桥前后连接，预示着江苏正在迈进一个崭新的时代。

设计规格：15m×6m×10m

Jiangsu Float

The Jiangsu Float features a Ruyi, the front of which is Suzhou Garden embodying Jiangsu's characteristics and the rear part of which is composed of the models of modern buildings, expressways, electronics, chemical industry and machinery. The front and the rear are connected by a modern bridge, denoting Jiangsu is entering into a brand-new age.

Design specification: 15m×6m×10m

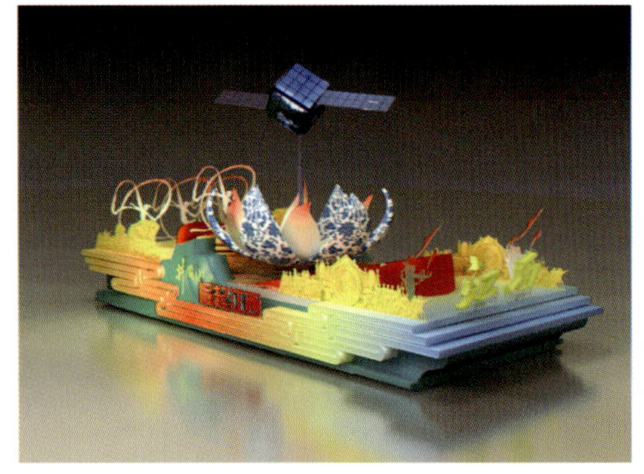

江西——崛起江西彩车

崛起江西彩车以喷薄而出的朝阳、如花蕾绽放的青花瓷球、人造卫星的两翼等为主体形象，从“红色江西、文化江西、现代江西”三个维度表现从传统走向现代、正在中部崛起的江西。

设计规格：15m×6m×10m

Jiangxi Float

The Jiangxi Float features the rising sun, blue and white porcelain ball and the two wings of an artificial satellite, indicating a rising Jiangxi from the traditional image to the modern image from three aspects of "Revolutionary Jiangxi, cultural Jiangxi and modern Jiangxi".

Design specification: 15m×6m×10m

山东——岱青海蓝彩车

岱青海蓝彩车以山东地图轮廓为主体形象，通过一只面向大海、振翅欲飞的雄鹰，祝愿祖国在经济全球化发展中腾飞、翱翔。

设计规格：15m×6m×10m

Shandong Float

The Shandong Float features the map outline of Shandong Province, expressing the wish of soaring in the development of economic globalization to our country with an image of a flying eagle facing the sea.

Design specification: 15m×6m×10m

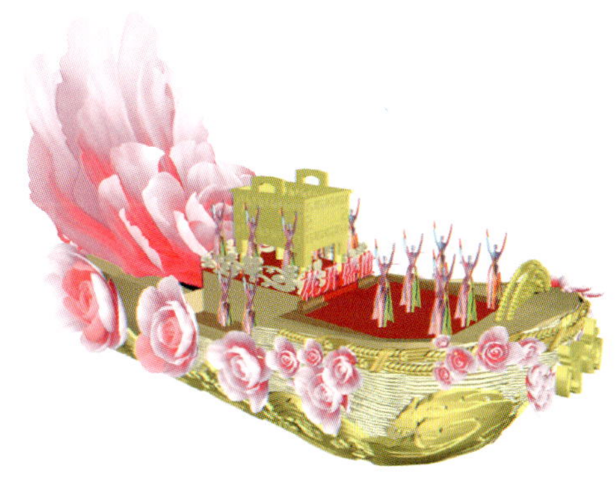

河南——花开盛世彩车

花开盛世彩车以牡丹为主体形象，以具有中原地域特色的青铜器元素、司母戊鼎变形而成的表演舞台、舞蹈表演等为主要元素，展现河南人民的豪迈和中原大地的雍容华贵。

设计规格：15m×6m×10m

Henan Float

The Henan Float features peony flowers, together with the elements like bronze wares with the Central China characteristics, stage transformed from the Simuwu Ding Cauldron and dance performance, displaying frankness and generosity of the Henan people and affluence of Central China.

Design specification: 15m×6m×10m

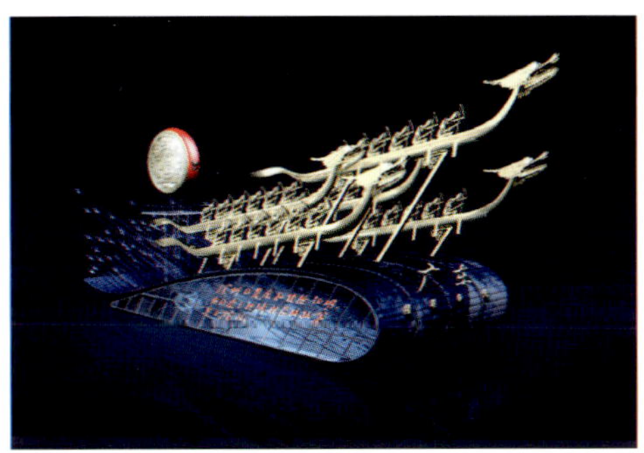

广东——领潮争先彩车

领潮争先彩车以琶洲会展中心为主体形象，屋顶动态翻转展现蓝色海浪或绿色岭南，五条“猛龙出海”表现未来发展方向和趋势，后方浪花托起双人击鼓寓意龙舟长风破浪。

设计规格：15m×6m×10m

Guangdong Float

The Guangdong Float features the Pazhou Conference and Exhibition Center. The dynamic overturn on the roof embodies blue sea waves or green Lingnan; the five dragons indicate the future development direction and trend; the two persons beating a drum with model of sea waves under their feet symbolize the dragon boat braving the wind and the waves.

Design specification: 15m×6m×10m

贵州——多彩贵州彩车

多彩贵州彩车以铜鼓为主体形象，舞台以苗绣图案为基础，苗族银角头饰造型形成一艘“巨轮”，象征贵州各族儿女在建设祖国的进程中乘风破浪、跨越发展。

设计规格：15m×6m×10m

Guizhou Float

The Guizhou Float features copper drum. The stage is shaped like a steamship of which the image is originated from the silver horn headgear of Miao ethnic group, with the background of patterns of the Miao embroidery, symbolizing that peoples from Guizhou brave the wind and waves to make breakthroughs in the process of development.

Design specification: 15m×6m×10m

最佳技术奖

Best Technology Award

牡丹花道具彩车

牡丹花道具彩车用国色天香的牡丹花，预示新中国成立六十周年国富民强的美好社会生活。

设计规格：11m×7m×6.2m

Peony Float

Peony flowers are used to foretell the good social life during the 60 years after the founding of the People's Republic of China.

Design specification: 11m×7m×6.2m

社会主义新农村彩车

社会主义新农村彩车以新农村现代化生产生活设施和农民业余文化生活作为主体形象，体现“生产发展、生活富裕、乡风文明、村容整洁、管理民主”的社会主义新农村建设主题。

设计规格：15m×7m×10m

New Countryside Construction Float

The New Countryside Construction Float features the modern production and living facilities and cultural activities in leisure time in the new socialist countryside, conveying the development themes of "developed production, easy life, civilized culture, neat village and democratic administration" in new socialist countryside.

Design specification: 15m×7m×10m

交通成就彩车

交通成就彩车以国产支线飞机、立体交通、铁路动车组和水上运输装备作为主体形象，重点展现交通运输业在规模、质量、技术装备水平方面发生的变化和取得的成就。

设计规格：15m×7m×10m

Transportation Development Float

The Transportation Development Float features domestic branch line airplane, three-dimensional transport, multiple units and marine shipping equipments, focusing on the changes happened and achievements made in size, quality, technology and equipment of the transportation industry.

Design specification: 15m×7m×10m

生态环保彩车

生态环保彩车以绿树、碧水、草地和绿叶状小船作为主体形象，体现人民群众充分享受生态文明建设的成果。

设计规格：15m×7m×10m

Environmental Protection Float

The Environmental Protection Float features green trees, clean water, grassplot and miniboat shaped like a leaf, embodying that the people enjoy the achievements made in the construction of ecological civilization.

Design specification: 15m×7m×10m

和谐社区彩车

和谐社区彩车以社区楼宇及社区服务中心、社区医院、居委会、文化游乐园等建筑作为主体形象，展现现代化城市社区“管理有序、服务完善、文明祥和”的社会生活面貌。

设计规格：15m×7m×10m

Harmonious Community Float

The Harmonious Community Float features the buildings of community service center, community hospital, community committee and cultural entertainment park, showing the living condition of the modern city communities——"orderly administration, improved service, cultivation and harmony".

Design specification: 15m×7m×10m

辽宁——振兴乐章彩车

振兴乐章彩车以展现老工业基地的发展为主题，以齿轮为主体形象，充分展现新中国成立60年来辽宁作为传统重工业基地在船舶、汽车、航空、高新技术产业等领域的重大成就与贡献。

设计规格：15m×6m×10m

Liaoning Float

The Liaoning Float features the development of the old industrial base and gear wheels, displaying the major achievements made in the areas of shipping, automobile, aeronautics and high-tech industry by Liaoning Province over the 60 years since the founding of New China.

Design specification: 15m×6m×10m

福建——扬帆海西彩车

扬帆海西彩车以祥云连卷细浪托起的彩船为主体形象，前部为活动舞台，惠安女翩翩起舞，中部为大型升降桅帆，船尾的七彩飘带象征海峡西岸势如虹的旺盛活力。

设计规格：15m×6m×10m

Fujian Float

The Fujian Float features a colorful boat pushed up by the waves shaped like the lucky clouds, the front of which is a stage on which the Hui'an Ladies dance, the middle part of which is a large-sized mast and the rear of which is a colorful ribbon which symbolizes the vitality of the west coast of the Straits.

Design specification: 15m×6m×10m

湖北——凤舞楚天彩车

凤舞楚天彩车以一只翱翔九天的凤为主体形象，代表湖北创新、激情、奉献的精神文化核心。

设计规格：15m×6m×10m

Hubei Float

The Hubei Float features a phoenix flying in the sky, symbolizing Hubei's major spiritual culture of innovation, passion and contribution.

Design specification: 15m×6m×10m

台湾——宝岛台湾彩车

宝岛台湾彩车以台湾岛形状为主体形象，阿里山、日月潭、传统歌仔戏、101大厦等展现台湾自然、文化以及现代化的都市风貌，以大海中升起的彩虹和两岸飞机对飞的造型表现两岸直接双向"三通"和两岸关系和平发展主题。

设计规格：15m×6m×10m

Taiwan Float

The Taiwan Float features a model of the Taiwan Island. The elements like the Ali Mountain, the Riyuetan Pool, the traditional Taiwanese Opera and the 101 Tower building reflect the natural scenery, culture and modernity of Taiwan. The rainbow rising from the sea and the airplanes of both the China mainland and Taiwan show the direct two-way "three communications" and the theme of peaceful development between the China mainland and Taiwan.

Design specification: 15m×6m×10m

绘就蓝图彩车

绘就蓝图彩车以"富强　民主　文明　和谐"立体文字、航船造型和红旗为主体形象，"未来号"航船象征着中华民族团结一心、勇往直前，在中国共产党的带领下实现宏伟目标的决心和信心。

设计规格：30m×7m×15m

Blueprint Float

The Blueprint Float features the three-dimensional words of "prosperity, democracy, civilization and harmony", model of boat and red flag. The Future boat symbolizes the Chinese nation work together to determinedly realize ambitious goals under the leadership of the Chinese Communist Party.

Design specification: 30m×7m×15m

最佳视觉奖

Best Vision Award

浴血奋斗彩车

浴血奋斗彩车以人民英雄纪念碑浮雕、艺术造型红旗作为主体形象，表现中国人民反对内外敌人、争取民族独立和人民自由幸福的光辉历程。

设计规格：30m×7m×15m

Bloody Struggle Float

The Bloody Struggle Float features the relief of the Monument to the People's Heroes and the artistic transformation of the Red Flag. It expresses the brilliant road of the Chinese people resisting both domestic and overseas enemies and fighting for national independence, freedom and happiness.

Design specification: 30m×7m×15m

艰苦创业彩车

艰苦创业彩车以工农兵雕像、解放牌汽车、三门峡、两弹一星、南京长江大桥作为主体形象，表现新中国建立后全国人民艰苦奋斗、奋发创业的精神风貌和丰硕成果。

设计规格：30m×7m×15m

Arduous Pioneering Float

The Arduous Pioneering Float features the sculptures of workers, peasants, soldiers, truck of the Liberation Brand, Sanmenxia, atomic bomb, hydrogen bomb and artificial satellite and Nanjing Yangtze River Bridge, conveying our diligent and pioneering national spirit and achievements made after the founding of PRC.

Design specification: 30m×7m×15m

体育成就彩车

体育成就彩车以运动场及领奖台为主体形象，自下而上分别展示全民健身、中华传统体育和竞技体育，象征着我国体育健儿不断冲击世界最高水平。

设计规格：15m×7m×10m

Sports Development Float

The Sports Development Float features playground and medals podium and displays national physical fitness, the traditional Chinese physical exercises and competitive sports, standing that Chinese athletes constantly break the world records.

Design specification: 15m×7m×10m

北京奥运彩车

北京奥运彩车以鸟巢和祥云火炬作为主体形象，奥运火炬象征着奥林匹克圣火永远不灭。车体显示屏播放北京奥运会精彩画面。

设计规格：25m×7m×15m

Beijing Olympics Float

The Beijing Olympics Float features the Bird Nest and the Torch. The Torch represents that the Olympic Flame will not turn off forever. The Float screen displays the wonderful Beijing Olympics.
Design specification: 25m×7m×15m

众志成城彩车

众志成城彩车以体现抗击自然灾害的圆雕、浮雕为主体形象，车身的冲锋舟造型，寓意全国人民同舟共济的精神。

设计规格：30m×7m×15m

Unity Float

The Unity Float features circular engraving and relief with the theme of resisting natural disasters. The Float is shaped like a boat, implying the national spirit of mutual help.
Design specification: 30m×7m×15m

黑龙江——龙腾盛世彩车

龙腾盛世彩车以龙为主体形象，结合金色的麦穗、展翅的丹顶鹤、劳作的磕头机、联合收割机等元素，充分展示了黑龙江新中国成立以来的发展与成就。

设计规格：15m×6m×10m

Heilongjiang Float

The Heilongjiang Float features dragon, together with elements like golden wheat ears, flying red-crowned crane, working pumpjack and combine harvester, displaying the achievements made by Heilongjiang since the founding of New China.
Design specification: 15m×6m×10m

广西——壮乡欢歌彩车

壮乡欢歌彩车以铜鼓、吊脚楼、绣球、梯田等为主体形象，体现广西的民族特色与风情。

设计规格：15m×6m×10m

Guangxi Float

The Guangxi Float features copper drum, wooden house projecting over water, ball made of strips of silk and terrace, indicating the characteristics of Guangxi ethnic group.
Design specification: 15m×6m×10m

西藏——和谐西藏彩车

和谐西藏彩车以西藏标志性的自然景观、人文景观为主体形象，以雪山、森林展现美丽神奇的自然景观；以布达拉宫展示保护民族优秀文化遗产。

设计规格：15m×6m×10m

Tibet Float

The Tibet Float features uniquely Tibetan natural scenery and cultural places of interest. The beautiful and miraculous natural scenery is showed through snow-capped mountains and forests. The Potala Palace is displayed to show the conservation of excellent cultural relics.

Design specification: 15m×6m×10m

甘肃——盛世华章彩车

盛世华章彩车以迎风招展的节日彩旗和飞天彩带为主体形象，借鉴敦煌壁画的构图方式，集中反映甘肃历史文化、现代建设等成就。

设计规格：15m×6m×10m

Gansu Float

The Gansu Float features the colorful fluttering festival flags and apsaras ribbons with reference to the composition of Dunhuang frescoes, reflecting the achievements in history, culture and modorn architecture.

Design specification: 15m×6m×10m

宁夏——塞上江南彩车

塞上江南彩车以回、汉男儿击鼓欢庆的巨型雕塑为主体形象，通过浪花、鱼、米、枸杞、明珠、能源化工基地等元素，象征宁夏的快速发展。

设计规格：15m×6m×10m

Ningxia Float

The Ningxia Float features a large-sized sculpture of men from the Hui and Han ethnic groups celebrating joyously by beating drums, with the elements of sea waves, fish, millet, medlar, bright pearl and energy and chemical bases symbolizing the rapid development of Ningxia.

Design specification: 15m×6m×10m

最佳造型奖

Best Model Award

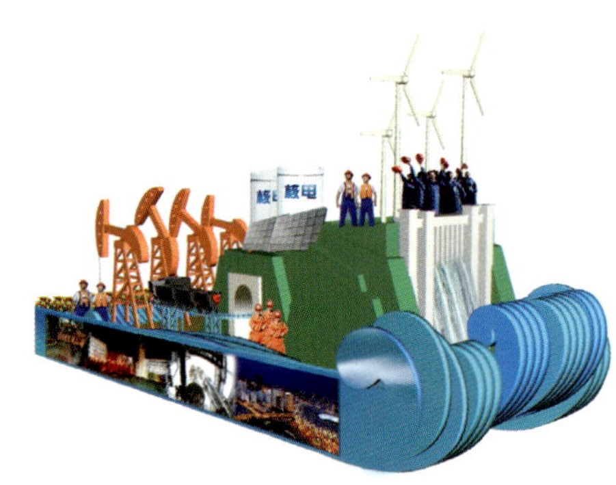

能源成就彩车

能源成就彩车以水利、煤炭、石油等传统能源和风力、核能、太阳能等清洁能源场景作为主体形象，彩车使用太阳能电池板即时发电供应，体现能源事业的可持续发展道路。

设计规格：15m×7m×10m

Energy Development Float

The Energy Development Float features the traditional energy sources like water conservancy, coal mine and petroleum and the clean energy sources like wind, nuclear and the sun. It uses solar energy battery panel for power supply, testifying the sustainable development of energy program.

Design specification: 15m×7m×10m

教育成就彩车

教育成就彩车以坐落在花岗岩基石上的打开的书本造型作为主体形象，车体显示屏播放我国在教育事业方面取得的各项成就，体现出我国教育事业"三个面向"、紧随时代发展。

设计规格：15m×7m×10m

Educational Development Float

The Educational Development Float features an open book on the base of granite. The Float screen displays our educational achievements, conveying that we stick to the educational guideline of "Gearing education to the needs of modernization, the world and the future" and keep abreast of the times.

Design specification: 15m×7m×10m

同一个世界彩车

同一个世界彩车以覆盖绿色植物的地球模型和我国外交方针"和平、发展、合作"立体文字为主体形象，象征着人类共同的家园欣欣向荣。

设计规格：15m×7m×10m

One World Float

The One World Float features a model of the Globe covered with green plants and dimensional words of "Peace, Development, Cooperation", symbolizing the flourishing of our common home——the Globe.

Design specification: 15m×7m×10m

团结奋进彩车

团结奋进彩车以立体红旗和向日葵花为主体形象，车体为船型，象征着中华民族在中国共产党的领导下同舟共济、共同前进。

设计规格：25m×7m×15m

Unity and Advancement Float

The Unity and Advancement Float features a dimensional red flag and sunflowers. The Float is shaped like a boat allegorizing that the Chinese nation work together to make progress under the leadership of the Chinese Communist Party.
Design specification: 25m×7m×15m

内蒙古——草原飞虹彩车

草原飞虹彩车以绿色草原为主体形象，融入骏马、蒙古族少女、六大支柱产业和风力发电等元素，展示发展成就。

设计规格：15m×6m×10m

Inner Mongolia Float

The Inner Mongolia Float features a green grassland integrated with the elements such as horse, young Mongolian ladies, six pillar industries and wind power to show the achievements.
Design specification: 15m×6m×10m

吉林——精彩吉林彩车

精彩吉林彩车以流线型高速动车组和数字"1"变化成飘动的红旗为主体形象，展示长白山脚下各族儿女载歌载舞、欢乐祥和。

设计规格：15m×6m×10m

Jilin Float

The Jilin Float features high-speed streamlined Multiple Units and wavy flag transformed from the shape of number "1", displaying the ethnic groups in the foot of the Changbai Mountain dance and sing happily and harmoniously.
Design specification: 15m×6m×10m

安徽——江淮和畅彩车

江淮和畅彩车以徽州古民居建筑元素——马头墙为主体形象，徽雕传统工艺多层彩云浮雕装饰在两侧，表达安徽改革开放、科学发展的时代心声。

设计规格：15m×6m×10m

Anhui Float

The Anhui Float features one of the elements in the traditional Huizhou folk house——Horse Head Wall, with the traditional craft of the Huizhou carving——layers of colorful cloud in the two sides, conveying Anhui people's wish to carry through the reform and opening policy and to develop scientifically.
Design specification: 15m×6m×10m

湖南——锦绣潇湘彩车

锦绣潇湘彩车以芙蓉花为主体形象，周围的流线造型象征三湘四水，车体前部球体将"韶山冲、红日和湖南地方特色花鸟"糅合在一起，表现湖南的繁荣昌盛以及对革命前辈的深切缅怀。

设计规格：15m×6m×10m

Hunan Float

The Hunan Float features cottonrose flowers, with its streamline framework standing for "three confluences and four rivers", the front ball of which integrates Shaoshanchong (the place where Chairman Mao was born and lived long), red sun and local flowers and birds together, indicating Hunan's prosperity and memory of the revolutionary father.

Design specification: 15m×6m×10m

云南——七彩云南彩车

七彩云南彩车以"彩云之南"中的彩色之云为主体形象，孔雀舞及哈尼梯田作为表现元素，"太平有象"的主体雕塑寓意"盛世太平"。

设计规格：15m×6m×10m

Yunnan Float

The Yunnan Float features colorful clouds, peacock dance and Hani Terrace, with "Tai Ping You Xiang" as its main sculpture signifying "prosperous and peaceful society".

Design specification: 15m×6m×10m

陕西——三秦新韵彩车

三秦新韵彩车以延安宝塔山、西安城墙、秦兵马俑、新舟60飞机、航天卫星测控、石化能源管道等为主体形象，表现历史文化的陕西和当代腾飞的陕西。

设计规格：15m×6m×10m

Shaanxi Float

The Shaanxi Float features the Pagoda Mountain in Yan'an, the city wall in Xi'an, the terracotta soldiers and horses of the Qin Dynasty, observation and control of aeronautic satellite and energy pipelines, showing both the traditional and the modern Shaanxi.

Design specification: 15m×6m×10m

最佳制作奖

Best Execution Award

工业成就彩车

工业成就彩车以工业轴承、光盘、汽车生产流水线及炼钢高炉作为主体形象，象征传统工业及现代信息技术的不断发展。

设计规格：15m×7m×10m

Industrial Development Float

The Industrial Development Float features mechanical bearings, disks, auto assembly line and steel-making blast furnace, standing for the continuous development of the traditional industry and the modern information technology.

Design specification: 15m×7m×10m

民主政治彩车

民主政治彩车以人民大会堂作为主体形象，车体显示屏播放人民代表大会、中国共产党领导下的多党合作、民族区域自治以及基层民主自治等制度建设的画面。

设计规格：15m×7m×10m

Democratic Politics Float

The Democratic Politics Float features the People's Great Hall. The Float screen displays the institutional construction like People's Congress, multi-party cooperation under the leadership of CPC, regional autonomy of ethnic minorities and democratic autonomy at the grassroots level.

Design specification: 15m×7m×10m

依法治国彩车

依法治国彩车以宪法单行本模型及其它法律文本模型为主体形象，体现中国特色社会主义法律体系已经基本建成。

设计规格：15m×7m×10m

Rule of Law Float

The Rule of Law Float features the models of the Constitution offprint and other law documents, telling people that the socialist law system with Chinese characteristics is basically established.

Design specification: 15m×7m×10m

人口卫生彩车

人口卫生彩车以公共卫生领域、预防与控制传染性疾病及计划生育工作标志为主体形象，车身造型为绿叶，象征着生命与希望。

设计规格：15m×7m×10m

Public Health Float

The Public Health Float features the public health area, contagious disease prevention and control and symbol of family planning. The Float is shaped like a leaf which stands for life and hope.

Design specification: 15m×7m×10m

河北——激情河北彩车

激情河北彩车以乘风破浪的巨轮、鲲鹏展翅的造型、现代城市建筑群、吴桥杂技等为主体形象，展示了河北人民在党中央领导下所取得的建设成就。

设计规格：15m×6m×10m

Hebei Float

The Hebei Float features a large steamship, flying roc, modern city buildings and Wuqiao acrobatics, displaying the achievements made by Hebei people under the leadership of CPC.

Design specification: 15m×6m×10m

海南——绿色海南彩车

绿色海南彩车以天涯海角、滑板风帆、水果鲜花、椰树等元素错落层叠为主体形象，诠释了一个绿意盎然、度假养生的多元化生态自然岛概念。

设计规格：15m×6m×10m

Hainan Float

The Hainan Float features elements of "the ends of the earth", slide board, sails, fruits, flowers and coconut tree, explaining the concept of a diversified ecological island full of green trees and suitable for holidays.

Design specification: 15m×6m×10m

四川——奋进四川彩车

奋进四川彩车以抗震群雕、"四川依然美丽"LED彩屏为主体形象，弘扬抗震救灾精神。

设计规格：15m×6m×10m

Sichuan Float

The Sichuan Float features the group carving themed with earthquake resistance and a LED color screen displaying that Sichuan is still beautiful, propagating the greatness of combating the earthquake and carrying out relief work.

Design specification: 15m×6m×10m

新疆——天山祝福彩车

天山祝福彩车以纳格拉木鼓、艾迪莱斯绸、瓜果等为主体形象，突出地域、民族特色和新型工业化建设成就。

设计规格：15m×6m×10m

Xinjiang Float

The Xinjiang Float features Nagela wooden drum, Aidilaisi silk and fruits, focusing on the regional and national characteristics and the achievements made in developing new industrialization.

Design specification: 15m×6m×10m

澳门——盛世莲花彩车

盛世莲花彩车以浪花和澳门城市建筑的典型元素为主体形象。车上的主体结构以澳门城区元素为主，辅以标志性新建筑，体现澳门发展史。

设计规格：15m×6m×10m

Macao Float

The Macao Float features sea waves and typical elements in the city buildings of Macao. The Float is mainly composed of urban elements, together with symbolic new buildings, indicating Macao's development path.

Design specification: 15m×6m×10m

繁花似锦彩车

繁花似锦彩车以满载花朵的巨型花篮为主体形象，在少年儿童的簇拥下，寓意着祖国的发展前景繁花似锦。

设计规格：10m×6m×10m

Flourishing Flowers Float

The Flourishing Flowers Float features large-sized baskets full of flowers and is surrounded by young children signifying the prospects of China are blooming.

Design specification: 10m×6m×10m

09

最佳团队奖

Best Team Prize

中国创新设计红星奖

China Red Star Design Award

中国电信上海研究院创新团队

中国电信上海研究院创新团队是一个横跨项目管理、市场研究、用户测试、产品设计、产品策划、信息技术等多专业、多领域的复合型创新团队。其成员以中国电信产品研发团队为主导，并吸收上海桥中设计咨询团队成员，为完成中国电信手机、网关等电子产品研发设计而组建。桥中与中国电信内部的设计团队共同承担产品的策划、调研、外观设计等工作。中国电信发挥其作为中国三大电信运营商之一的平台优势，整合产业链的各个环节，实现概念的产品化。历时一年，经过对30多位盲人的调研，一款以人文关怀为出发点、让盲人走进3G时代的手机——九宫盲人手机诞生了。通过此次合作，该团队由产业链上端牵头，整合各环节研发创新力量并管理整个产品开发流程的"从创新到产品"一体化的模式是值得借鉴的。

Shanghai Research Institute of China Telecom Co., Ltd. Innovation Team

Shanghai Research Institute of China Telecom Co., Ltd. Innovation Team is a versatile team mastering multidisciplinary expertise such as project management, market research, user testing, product design, product engineering, information technology, etc. The team is assembled for designing and developing China telecom electronic products like cell phone and gateway. The members coming from products design division of China Telecom Co. Ltd. and China Bridge International jointly charge of product engineering, researching, outlook designing, etc. Making full advantage of its resources as one of the three main telecom operators in China, China Telecom Co. Ltd. integrates every link of industry chain and brings design concept into reality.

After one year of endeavor and research on 30 blind people, 9 spots "mobile global eye" came into being. Through this cooperation with China Bridge International, the mode of integrating every innovative force and managing the whole product design process is worth learning.

09

最佳新人奖

Best New Designer Prize

中国创新设计红星奖
China Red Star Design Award

羊文军

羊文军出生于1985年12月，毕业于浙江大学工业设计专业。大学期间参加浙大创新团队，与团队成员共同创作的概念作品曾获德国红点概念奖。2007年毕业后，加入中兴通讯担任工业设计师，参与公司系统产品的PI规划；成功完成B600E、多媒体终端、9000系列和M6000系列等多个项目，以其独特的风格赢得内部客户和市场的信任与好评。
作为"睿智"——ZET中兴T8000高端路由器的造型设计负责人，他从外观设计、材质选择、人机原理等方面都进行了详细的调研，设计的方案对系统产品进行了一次视觉革命。T8000高端路由器充分体现了设备的专业感和品质感，该产品荣获了2009中国创新设计红星奖金奖。

设计感言："设计要考虑各种制约因素，充分发挥自己的创造力，寻找最佳的视觉语言向大家表达自己的想法。"

Wenjun Yang

Wenjun Yang was born in December 1985, graduated from Zhejiang University, majoring in industrial design. He has once won red dot: design concept in cooperation with other team members. After graduation in 2007, he joined in ZTE Corporation as an industrial designer, participating in PI programming and successfully accomplishing projects like B600E, multimedia terminal, 9000 series, M6000 series, etc. His unique design style won high praise and trust from inner customers.
As the designer in-charge of "Sagacity"-ZTE T8000 High-end Router, Wenjun Yang did thorough investigation in appearance design, material selection, ergonomics, etc., and made a system product plan which can be called a visual revolution. Highly professional T8000 demonstrates outstanding quality and won Gold Prize of 2009 China Red Star Design Award.

Design Idea: A good designer need to think over every possible restraint, make full advantage of his own strongpoint and seek the best visual language to express his design ideas.

红星奖
Excellent Prize
中国创新设计红星奖
China Red Star Design Award
09

DHOME陶瓷产品设计工作室

Purity青花酒瓶

本产品朴素如玉的外表和幽雅的景德镇青花手绘工艺融合了中西文化。盖上瓶盖，酒瓶体现出西式的简约和中式的素雅；打开则是纯粹的中式味道，瓶内的青花让人眼前一亮。瓶体表面使用亚光釉，便于手持不易滑落，瓶盖还可用作酒杯。

DHOME CERAMIC PRODUCT DESIGN WORKSHOP

Blue and White Bottle

Through a simple appearance and Jingdezhen's elegant hand painting craftwork, the design exposits an integration of Chinese and Western Culture. With the cap, its appearance shows a perfect combination of western simplicity and Chinese elegance; while without the cap, the inside shows an absolute Chinese style, which surprises us by the marvelous blue-and-white pattern. The outside of the body is matte-glazed, making it more comfortable and safe to handle; in addition, the cap can be used as a goblet.

DHOME陶瓷产品设计工作室

H&D视力表钟

通常人们看书或看电视久了很容易近视，本设计使人们通过看时间来控制阅读时间，并可测量视力提醒用户保护眼睛。

DHOME CERAMIC PRODUCT DESIGN WORKSHOP

Test-Chart Clock

People always easily get myopia by watching TV or reading for a long time. With this product, users can control the time of TV-watching or reading by using this clock to check time. It also reminds users to take care of their eyes by measuring eyesight regularly.

北京博蓝士科技有限公司

M50高精磨床

M50高精磨床的设计以直线型为主，辅以圆弧转角，体现简洁、人性化的形态。创新的开门拉手设计，使产品操作简便。在人机方面，悬臂操作箱使操作者无需长距离或频繁移动便可从工作台位置清楚了解加工数据。

IDCdesign

M50 High Precision Grinding Machine

The M50 series 1363 type with the design of linear and circular arc angle creates a concise and humanized form. With the design of a main linear pattern and aided by a circular arc angle, the M50 series 1363 type reflects a concise and humanized form. An innovatory design of the handle simplifies operation. Concerning ergonomics, cantilever operation box enables operators to know the processing data clearly just at the workbench without distant and frequent moving.

TCL多媒体科技控股有限公司

TCL Multimedia Technology Holdings Ltd.

TCL X10液晶电视

TCL X10 LCD TV

TCL新一代液晶电视X10总厚度仅45毫米，设有前置音箱。推拉抽屉式的连接器方便美观，底部按键触感强，使用便捷。托盘式底座可放置其他多媒体设备，而且它采用注塑高光材质，莹泽流畅。

The new generation LCD TV X10 of TCL is only 45mm thick with an integrated speaker set forward, while the exquisite cable management system is set at the rear. Connector drawer in the front is a push–push mechanism, which is not only convenient to use and pleasing to eyes but also owns strong–touch–sense keys. Tactile pattern on the bottom guides users where to press. Tray–like base provides space for media devices.

TCL多媒体科技控股有限公司

TCL Multimedia Technology Holdings Ltd.

TCL C10液晶电视

TCL C10 LCD TV

C10液晶电视拥有超薄"刀锋"面板和优雅的前后过渡；集成USB功能，可连接摄像机等多媒体设备；不锈钢薄板Logo给人高品质的视觉冲击。金属材质的高光黑前壳，丝绸质感机身都是引领时尚潮流的亮点。

New generation LCD TV C10 is an integration of super slim "Blade" bezel, elegant front–to–back transition and USB port for camera. Connector drawer is a push–push mechanism. Stainless steel sheet of TCL Logo slides with a high–quality feeling. Finished black anodized aluminum front panel and the silk–tactile impression are all high lights leading the fashion trend.

齐思工业设计咨询（上海）有限公司

TEAMS Design Consulting Co., Ltd.

爱特那卫浴套装

Etna

爱特那是一套Spirella的浴室系列产品，包括香皂盒、牙杯、洗手液容器，便桶刷等。外形上，设计灵感源自天然不对称的鹅卵石造型，使用户贴近大自然。它的设计语言植根于日本著名的园艺手法"枯山水"，考虑的不仅是美观，还有使用是否方便。

Etna is a collection of bath products created by Spirella, including soap boxes, tooth mug, liquid soap containers, and bucket brush and so on. The appearance is inspired by the organic and asymmetrical formations found in cobble stones, reflecting the beauty of nature. Etna's design language reflects the calming and soothing elements rooting in famous Japanese garden art known as "Karesansui", serving the purposes of form and function.

齐思工业设计咨询（上海）有限公司

GBM 6RE、10RE、13 RE 专业电钻系列

博世GBM专业钻头系列产品的把手非常适合亚洲人手的大小。机身紧实、轻巧，适合使用者在狭小空间内或恶劣环境中工作。创新的皮带夹便于使用者在脚手架上攀爬，从而提高了灵活性和安全性。

TEAMS Design Consulting Co., Ltd.

GBM 6RE, 10RE, 13 RE Professional

BOSCH's GBM 6RE, 10RE and 13RE drills for professional construction workers are designed specifically for Asian market. For example, the ergonomic handle is suitable for smaller hands, small size also allows users to reach in narrow spaces, and a sturdy design is able to withstand the roughest environments. The product is equipped with an innovative belt clip allowing the worker to use both hands for bracing while walking on scaffolds.

齐思工业设计咨询（上海）有限公司

GWS 7–100专业角磨机

博世GWS 7–100的设计在性能和设计上均有显著的提高，尤其体现在对人机性和安全性的处理上。它的把手方便手较小的使用者抓得更牢固、更舒适。前部齿轮的外壳齿轮箱可轻松旋转90度，以适合习惯左手或右手的使用者。

TEAMS Design Consulting Co., Ltd.

GWS 7-100 Professional

The brand new GWS 7–100 is significantly improved in both performance and design, especially in terms of ergonomics and security. The circumference of the handle allows users with smaller hands to grasp firmly and comfortably. The front gear housing can be rotated in 90° , making it suitable for both right hand and left hand.

齐思工业设计咨询（上海）有限公司

GDR 14.4伏/18伏 锂电池专业电钻

博世GDR 14.4伏/18伏锂电池专业电钻，以屋顶为主要施工对象。为减少或防止使用过程中工具的跌落和滚动，它在设计细节作了特别的处理：它的头部较为短小，把手设计得非常人性化，即使在狭小空间使用，握感也非常舒服。

TEAMS Design Consulting Co., Ltd.

GDR 14.4V-Li, 18V-Li Professional

BOSCH's GDR 14.4 and 18 Volt Li–Ion impact drills are designed primarily for Asian market. Often used on roof–tops, the design focuses on details to minimize or prevent the tool from dropping, rolling, or tumbling. The short head makes it handy even when working in narrow spaces while the ergonomic handle allows a good grip.

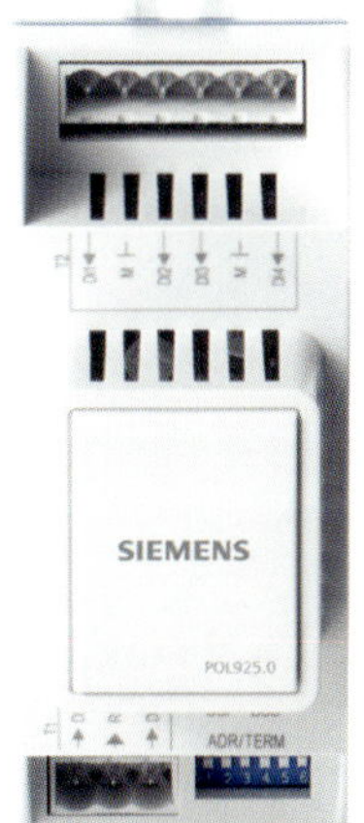

齐思工业设计咨询（上海）有限公司

中央暖通控制系统

Climatix是一个模块化的暖通控制系统，适用于所有加热和制冷的应用模式，在–40℃~70℃的极限环境仍可以正常运行。这个方便用户操作的应用系统，通过准确、有效的运转，降低了用户的支出。

TEAMS Design Consulting Co., Ltd.

Climatix

Climatix is a modular controller range HVAC system designed for any type of heating or cooling application. It is suitable for extreme operation conditions from –40℃ to 70℃. With its intelligent and proven applications, Climatix offers accurate and efficient operations while reducing total cost.

鞍山森远路桥股份有限公司

SY4500沥青路面热再生重铺机组

SY4500沥青路面热再生重铺机组是一种新型路面大修设备，可100%利用原有路面材料，施工时不产生废弃料，不占用土地，不会产生噪声和有害气体，不用封闭交通。与传统施工方法相比，施工周期可缩短70%以上。

AnShan SenYuan Road and Bridge Co., Ltd.

SY4500 Asphalt Road Surface Hot-regenerating Re-paving Unit

The SY4500 asphalt road surface hot–regenerating re–paving unit is a new heavy repair of road equipment, consisting of two road surface heating machines, one heating miller machine and one heating remixing machine. It is substantially used in high–grade highway area, and does repair work continuously such as heating on the spot, turning and fluffing road, adding regenerate reagent and so on.

北京爱慕内衣有限公司

运动支撑内衣AM62311–11

运动支撑内衣主要针对各种球类运动和各种有氧、无氧健身训练所需，关键设计点在于研究运动过程中的肌肉支撑及损伤、肌肉能耗，关注女性运动过程中的胸部支撑和保护。

Beijing Aimer Lingerie Co., Ltd.

Sport Corsage

Sports underwear is especially designed for ball sports and aerobic or anaerobic exercises. On the basis of research on muscle group support, injury, and energy consumption, the design focuses on support and protection of women's chests.

北京格物创道科技发明有限公司

BEIJING CREIWAY INVENT&DESIGN Co., Ltd.

T1988宽屏液晶电视

T1988 LCD TV

T1988宽屏液晶电视前壳采用分色设计，能形成多种时尚靓丽的配色方案，满足消费者的个性需求。独特的电源设计为整体造型添加亮点，带来全新的视觉体验。19寸机身针对都市家居时尚生活设计，可壁挂或加底座。

T1988 widescreen LCD TV is a product designed for urban home lifestyle with smooth elegant shapes. The former shell of the TV adopts separation color design, possessing a wide range of products designed to fashion the beautiful color scheme to meet different consumer demands. Unique power supply design adds highlights to for the overall design, creating a whole new visual experience. The 19-inch machine body can be hung on the wall or added with a new base.

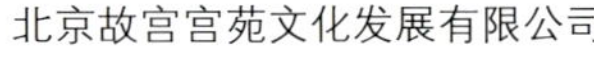

北京故宫宫苑文化发展有限公司

Beijing Imperial Court Cultural Development Company Ltd.

圆雕《清·乾隆皇帝阅兵像》

Emperor Qianlong Inspecting The Troops

《清·乾隆皇帝阅兵像》是在著名清代宫廷西洋画师郎世宁（意大利人）的巨幅绘画《乾隆戎装大阅图》基础上参考大量清宫档案及实物创作而成。采用冷瓷、合金、水钻等多种材质，经雕塑、制模、铸压等数十道工序完成。

The demonstrated sculpture artwork "Emperor Qianlong Inspecting the Troops" was reproduced and redesigned on the basis of the imperial painting by Lang Shining (Giuseppe Castiglione), basing on numerous data and collections of Imperial Palace Treasures. It adopts many materials such as cold porcelain, alloy, as well as dozens of procedures such as sculpture, model-making to finish.

北京华旗资讯数码科技有限公司

Beijing Huaqi Information Digital Technology Co., Ltd.

电子书R6

E-book R6

这款电子书外观简洁，翻页的五向按键居中并凸起，能够进行盲操作。它还具备上网及阅读图片的功能。独有的电纸书E-INK屏幕能在阅读中有效保护眼睛，此外它还十分环保，既节约纸资源又具有低能耗的特点。

Being concise in appearance and available for blind operation, the E-book is also entitled to connect with internet. The unique screen is effective for the eyes-protection. In addition, it is also environment-friendly, saving both paper and power.

北京华旗资讯数码科技有限公司

Beijing Huaqi Information Digital Technology Co., Ltd.

数码录音笔(专业型)

High-End Digital Voice Recorder

数码录音笔（专业型）机身采用金属磨砂材质与多触点防滑按键，不易磨损，手感舒适。第二代AGC增益控制模块，有效防止录音爆音及电平过低。内置时钟芯片，配合第二代录音时间戳技术，可在本机直接察看录音文件时间。

The digital recording pen has a metal frosting fuselage, which is exquisite, fashionable, comfortable and durable. The synchronous audio monitor of recording design adjusts recording parameter at any time and guarantees best recording effect. The built-in clock chip together with the second generation recording time-stamp technology supports the machine to check recording time directly.

北京华新意创工业设计有限公司

IDEAS DESIGN Co., Ltd.China

方正VEFOUND“文房”3G电子书

Founder E-book

方正电子书阅读器是针对都市白领设计的信息获取终端，基于中国移动的网络平台，能同时具备网络媒介和传统纸质书籍的优势。符合人体工程学的外观设计，保证使用者在操作的同时，其视觉重心集中在屏幕上。

Founder e-book is an information acquirement terminal designed for white collars in metropolitan cities. Though the network of China mobile, it acts as an internet media as well as a traditional book. The outlook adopts ergonomics, allowing users' attention focused on the screen.

北京嘉寓门窗幕墙股份有限公司

Beijing JIAYU Door,Window and Curtain Wall Joint-Stock Co., Ltd.

双层智能呼吸窗1系

Double-decker Smart Respiratory Window

双层智能呼吸窗由内、外两层窗户组合而成，同时在中间层设置遮阳装置，内外层窗之间形成一个相对封闭的空间，集保温、隔热、遮阳、通风、隔声多种功能于一身。

Double-decker smart respiratory window is newly developed with multifunctions of heat-insulation, shading, ventilation, sound-proof etc. It is assembled by inner and outer layers with shading equipment in the interlayer, forming a sealed space.

北京蓝海洋文化传播有限公司

御米油(罂粟籽油)系列包装

该产品是从罂粟籽中提取的不含吗啡的高档食用油，以及以罂粟籽油为原料加工而成的营养胶囊。在设计中，尽可能少地使用色彩，使油本身的颜色更加突出；纹样给人以严谨、科学之感；油瓶造型的灵感来自人体的优美曲线和葫芦的形状。

Beijing Blue Ocean Culture Communication Co., Ltd.

Poppyseed Oil & Soft Capsule Package Design

The package design intends to represent the product's complex identification: The product is high-grade, morphine free, with edible oil extracted from poppy seeds. The color of the package is kept simple to reflect the original oil color, while traditional patterns announce authority and classic, and the shape of the container gains inspiration from the shape of Hyacinth.

北京蓝威科瑞科技有限公司

猎鹰监控系统

"猎鹰"是一款装备多功能，军、警用，高科技全天候下进行野外侦查、蹲点、抓捕、搜救的多传感器成像望远设备。其设计非常注重满足军警苛刻的使用环境要求。外观既有军用品粗犷的风格，又不失丰富的细节。

Beijing Lever Create Co., Ltd.

Falcon

Falcon is a multi-sensor imaging device, attaching great importance to military harsh environmental requirements. It not only has a military style but also owns abundant details.

北京蓝威科瑞科技有限公司

狩猎者红外夜视镜

"狩猎者"是一款小型军警用简易红外夜视镜设备，可使军、警人员在无光环境下执行罪犯搜捕、监视等任务。此款产品轻巧、便于携带。镁铝合金的金属壳体牢固、耐用。

Beijing Lever Create Co., Ltd.

Hunter

Archer-Hunter is a simple and small infrared night-vision goggles, an equipment for military and police use enabling the military and police officers to carry out many duties in matt environment, such as offender arrests, surveillance. Its small size, light weight characteristics make this product very portable. Simple functions make the operation very easy.

北京品物堂产品设计有限公司

桌面3G路由G3R01

随着3G技术在中国的普及，设计师们也开始将3G路由器的形象赋予"新"的概念。插花的花瓶和网络路由器被完美地结合在了一起，使本身中性的网络产品具有了感情色彩，并起到了装饰作用。

PER Design

3G Desktop Router

With the increasing popularity of 3G technology in China communication industry, the designers think it should embody the "new" concept. Vase, net router are two completely different things, but their perfect integration makes the neutral internet product own an emotional coloring. In addition to its network function, it also can be used as a decoration.

北京品物堂产品设计有限公司

OBU电子高速计费器

OBU电子高速计费器是用在高速路自动收费的产品，可实现不停车收费计费，不仅节约了用户的时间，也减少了高速系统的人力资源的投入。外形简洁，基调与车内装饰十分协调，插卡和读取数据简单轻松。

PER Design

OBU Electronic Highway Toll Collection

This product is used in automatic highway toll collection system. Vehicles with this installment do not need to stop to pay the toll, as it can communicate with a roadside collection station. The LED screen hiding below the black translucent material can't be seen without lighting, making the product more holistic.

北京品物堂产品设计有限公司

家庭无线网关G3R09

圆润？光滑？简约？气质？……一直找不出最恰当的词语来描述设计师对家庭无线网关的理解，总觉得它永远比我们想象的要丰富。它像跳跃的精灵根植于我们思想的角落，激发着我们对生活的热爱。

PER Design

Wireless Home Gateway

Smooth? Glossy? Concise? Graceful?…no proper words could be found to describe how designers appreciate this product. It is always like something that more than we can imagine. It grows seemingly around certain edges of our mind, waiting to meet someone who loves the life.

北京品物堂产品设计有限公司

PER Design

无线多媒体信息交互终端机

Wireless Multimedia Terminal For Information Exchange

无线多媒体信息交互终端机用简单的方和圆为基本元素，巧妙搭配成各种形态，颜色上采用灰色基调，自然、大方，配上细腻的表面处理，使产品看上去具有很高的品质和科技含量。

Wireless multimedia information exchange terminal takes square and circle as the basic elements, and the whole shape is the perfect match of the two elements. As for coloring, grey is a keynote——natural and generous. The exquisite surface treatment gives it high–level and technological appearance.

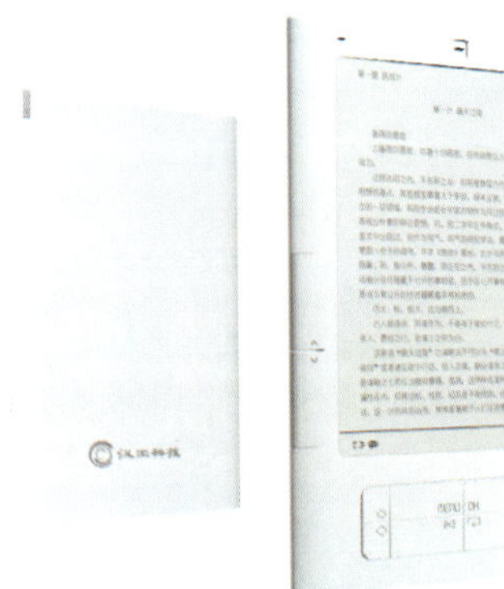

北京品物堂产品设计有限公司

PER Design

电子阅读器Walkreader

E-book

在互联网信息时代，电子阅读器已成为新的读书方式。该设计小巧、轻薄、节省空间、易于携带，充电一次可以工作200小时以上，并且有着书本不可替代的容量优势。

In this internet time, people gain information initiatively or passively through channels such as internet, TV, film etc. However, a new reading carrier Electronic Reader is becoming popular. The product is so smart, thin, space saving and portable. It can work at least 200 hours after one recharge.

北京品物堂产品设计有限公司

PER Design

钥匙扣 CKEYS

Keyclip

通常，找出掉在包底的钥匙会花很多时间。这个钥匙扣就是为解决该问题而设计的。带有弹性的材料，可将钥匙扣别在包的内壁口袋上或者中间的夹层上，防止其掉到包底。

Sometimes, looking for the key at the bottom of the bag takes a lot of time. This keyclip is designed to solve this problem. Bouncy material, complete smooth shape and nice function allow the keyclip to attach to the inner bag or an interlayer, preventing it dropping into some corners that are not easy to find.

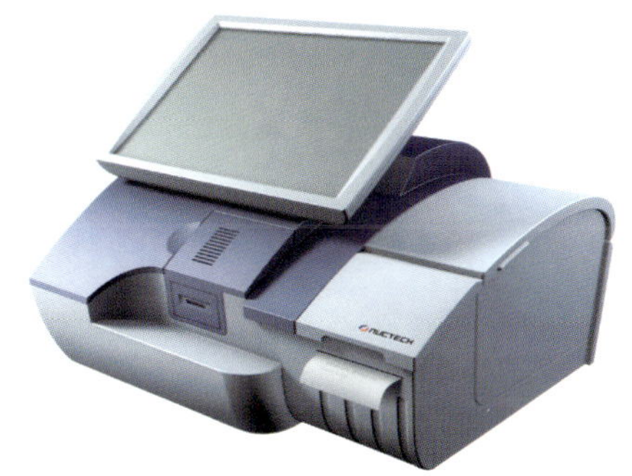

北京品物堂产品设计有限公司

PER Design

便携式爆炸物检查仪

Portable Explosive Detector

该产品用于机场、海关，用来检测爆炸物和毒品，具有针对爆炸物的痕量分析的功能，可快速、准确判断痕量爆炸物的存在，并分析出爆炸物的种类和威胁程度。设计简洁，操作舒适。

This product is used to detect explosives and drugs at airports, quays, border ports and crowded places, etc. It has the capability of analyzing the explosive trace amount and the types and threat level quickly, compact and comfortable to use. Concise design, yet comfortable operation.

北京品物堂产品设计有限公司

PER Design

个人基因分析系统

Gene Analysis System

该产品是对个人基因进行分析的仪器，通过检测身体的遗传基因，对各类遗传病进行有效的预防。设计简洁大方，富有现代气息。

This product is designed for gene expression analysis, i.e. checking the body information and preventing disease effectively. This gene analysis system appears simple, but it is of good quality and more modern than traditional medical equipments.

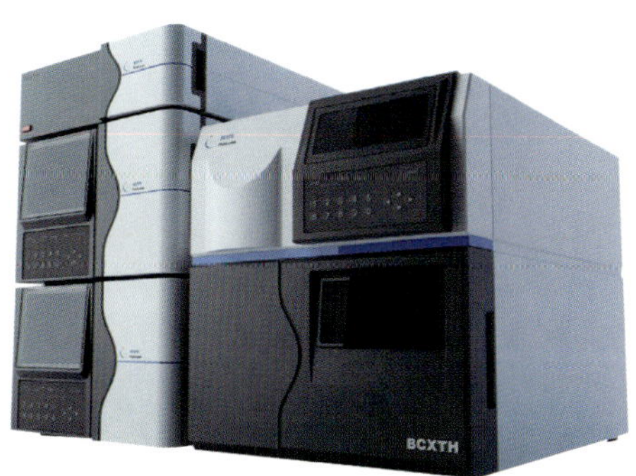

北京品物堂产品设计有限公司

PER Design

液相色谱仪AJ7302

Liquid Chromatograph

这款液相色谱仪，灰白色的基调符合实验室仪器特有的审美标准，材料选择和表面处理让产品质感十足。每个模块都可单独使用，组合后又井然有序；S形线条，仿佛能看到检测液体向下流动的形态。

This is a product for liquid chromatographs. White－grey motif complies with the aesthetical standard of an apparatus in laboratories. Modular design is most characteristic in this design. S line seems like the flowing status of fluid being tested.

北京曲美家具集团有限公司

C3椅

弯曲的椅背与后腿合二为一，前腿与座面连接处采用"三碰肩"技术，突破了传统的直角结合形式，且坐感舒适。实木单板胶合弯曲，提高了木材的利用率。整体采用枫木色，时尚、淡雅、大方。

BEIJING QUMEI FURNITURE GROUP CORP., LTD.

C3 Chair

The bent chair-back is joined with two back legs, and the front legs and seat form an angle perfect for people to relax. Bent solid wooden veneers enhance utilization rate. Maple color adds a sense of fashion and elegance to this chair.

北京水童星科技发展有限公司

高尔夫实球击打辅助训练系统

该系统通过减少球体重量与增大飞行阻力的双重措施，使人们可以在小空间甚至客厅中使用鹰球进行木杆与铁杆的实球击打训练，大大降低该项运动的成本。

Beijing Savater Technology Development Co., Ltd.

Confined Space Golf Stroke (CSGS) Training System

The CSGS Training System reduces the weight of golf and increases its flying air resistance, making it possible for people to practice stroke training with wooden or iron shaft in a 40m×3m×4m room or even in the living room. Moreover, it can be used anytime and anywhere, helping people improve their skill much faster.

北京探路者户外用品股份有限公司

明斯克40升背包

明斯克40升背包采用探路者Convenient Trek技术，颠覆传统概念，从背部开孔。使用者休息时，习惯于把背包的面部接触地面；坐在背包的背负上，可同时打开背负中间的双拉链，开口直通背包大身，便于取包内物品。

Beijing Toreading Camping Equipment Co., Ltd.

MINSK Backpack 40L

MINSK Backpack 40L adopts the "Convenient Trek" technology. Subverting the traditional concept, the product has a door opening from the back. Basing on years of design and outdoor experience, the design is very practical.

北京探路者户外用品股份有限公司

尼尔徒步鞋

这是一双带有BOA松紧系统的多功能户外运动鞋，与传统的鞋带系统不同，BOA系统任何时刻都可进行微小而舒适的调整，不会变松，也不需要重新扎系。鞋底的设计可同时满足徒步时的稳健抓地和溯溪时的平稳防滑。

Beijing Toreading Camping Equipment Co., Ltd.

Neil Trekking Shoes

It is a pair of multi-functional shoes with a BOA elasticity system. As tightness can be finely tuned by the system, it can't loosen so users don't need to lace again and again. The soles have different design patterns to make the shoes grip the ground steadily while walking and anti-skid while tracing in the water.

北京探路者户外用品股份有限公司

帝力可加热冲锋衣

该产品采用电子控制的恒温加热系统，在最需保护的胸口和背部设置加热点，需要时开启加热系统，30秒以内便能让人感觉到温暖，加热时长达2.5~11小时。此外还设有一个安全的控制系统，温度控制在36℃~65℃之间。

Beijing Toreading Camping Equipment Co., Ltd.

Dili Heated Jacket

This heated jacket is different from current ones, as it utilizes and sews the carbon fiber thermal heating elements into the liner jacket and works with the rechargeable Li–Polymer Battery & Multi–setting temperature controller to create the Far Infrared Ray (FIR) heat. It is specialized in the low power consumption so it keeps the hours of warmth efficiency.

北京探路者户外用品股份有限公司

马里可加热冲锋裤

这款马里可加热冲锋裤使用电子控制的恒温加热系统，针对最需要保护的膝盖设置加热点。在需要的时候，开启加热系统，整个系统在30秒以内，能让人感觉到温暖的包围，加热时间长达2.5~11小时。并且整个系统有一个安全的控制系统，温度控制在36℃~65℃之间。

Beijing Toreading Camping Equipment Co., Ltd.

Mali Heated Pants

This pair of heated pants is different with normal pants as it utilizes and sews the carbon fiber thermal heating elements into the fabric of the pants and works with the rechargeable Li–Polymer Battery & Multi–setting temperature controller to create the Far Infrared Ray (FIR) heat. It is specialized in the low power consumption so it keeps long hours of warmth efficiency.

北京突破电气有限公司

突破防水系列插座

这是一种防护等级达到IP55的民用移动防水插座，能够实现防溅水、防淋水和防喷水。绿色环保的防水材料TPE，完全符合欧盟RoHS标准。工艺简洁可靠，壳体棱角的槽状设计在传达一种强力、可靠感觉的同时，追求一种原生态仿生美感。

BEIJING TOP ELECTRIC Co., Ltd.

TOP Waterproof Socket

TOP Waterproof Socket is a civilian mobile waterproof socket whose protection level reaches IP55, which means effectively preventing water-splashing, water-trickling and spraying. All these make it the very first try for civilian mobile sockets. The adoption of environmental water-proof material, TPE, totally conforms to RoHs standard of EU.

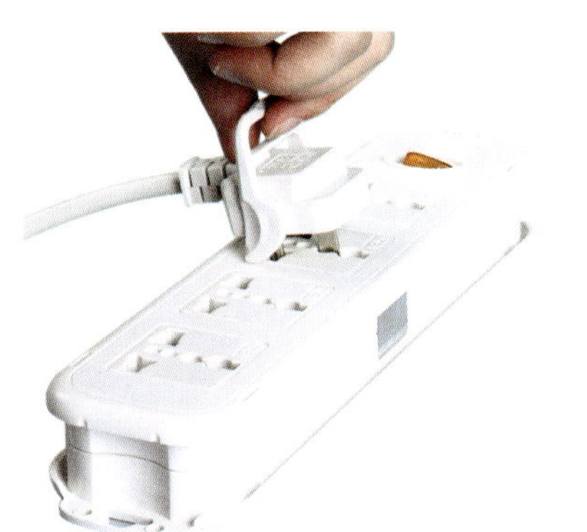

北京突破电气有限公司

突破易拉宝插头

这是一种带助力装置的插头产品。它利用杠杆原理，在8毫米的行程中轻松将插头从插孔中取出。在插头拔出的过程中可以单手操作，解放另一只手。成本较低。

BEIJING TOP ELECTRIC Co., Ltd.

TOP EASY PLUG

TOP Easy Plug is a kind of socket with assistant devices, using leverage principle to easily remove the plug from the socket hole during an 8mm trip. It is available for single-hand operation during plugging and pulling. Aesthetic shell edges are with trough-like design and look full of muscle sense which can convey a strong, reliable feeling, meanwhile, giving the product an elegant shape.

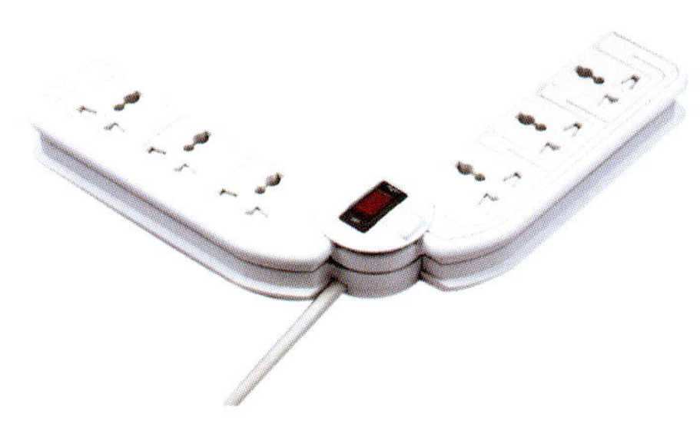

北京突破电气有限公司

突破旋转变形插座

这是一种可旋转变形的民用移动插座。插座可旋转，用户可以根据使用环境在水平方向从0°至180°之间自由选择使用角度，使用更加安全和方便；造型新颖，性价比较高。产品打破传统插座造型，融入旋转变形概念，个性鲜明。

BEIJING TOP ELECTRIC Co., Ltd.

TOP Rotary Socket

TOP Rotary Socket is a civilian mobile socket available for rotation. This type of socket is capable of rotation and user can select angles ranging from 0° to 180° according to using environment, making the operation much safer and more convenient. Also it can be used conveniently as dual socket or single one according to user's requirements.

北京心觉工业设计有限责任公司

Beijing Sense Perception Industrial Design Co., Ltd.

园林取土器

Post Hole Digger

产品造型简洁大方，曲线造型让产品更为丰富，设计感十足。把手线条流畅大气，严谨的设计分析使把手更为人性化，舒适感加强。铲头的“力变向”让使用者能更省力地使用。

The product shape is simple and generous, breaking the former appearance of the original products. Curve shape endows the product a full sense of fashion. The handle's shape line is smooth, and the rigorous analysis enables the handle to be more user–friendly, thus enhancing comfort.

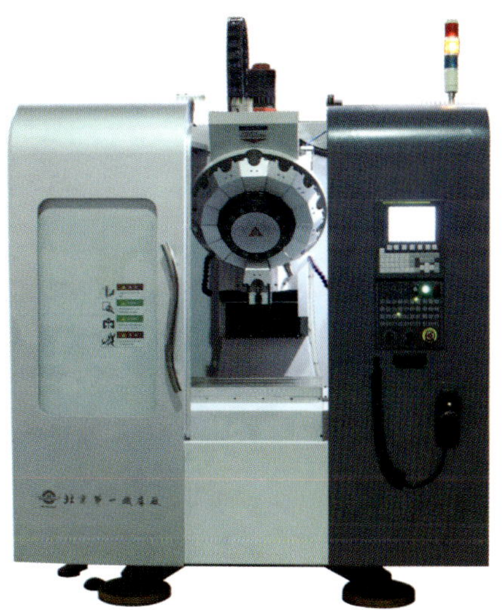

北京信息科技大学机电工程学院

Mechanical & Electrical Engineering School of Beijing Information Science & Technology University

ZH5120D立式钻削加工中心

ZH5120D Vertical Drilling Machining Center

ZH5120D立式钻削加工中心取消了原来防护罩开门上的横梁，既方便较重工件的装卸，也不会发生操作人员更换刀具、拿取工件时头部撞横梁的现象。将原来的联动双开门改为单开门形式，方便操作人员观察工件加工情况。

ZH5120D vertical drilling machining center changes the former door's structure and operation way. It removes the cross beam on the door of the shield. This design not only makes it easier to load and unload heavy work pieces, but also prevents workers from hurting their heads when changing cutters and taking work pieces.

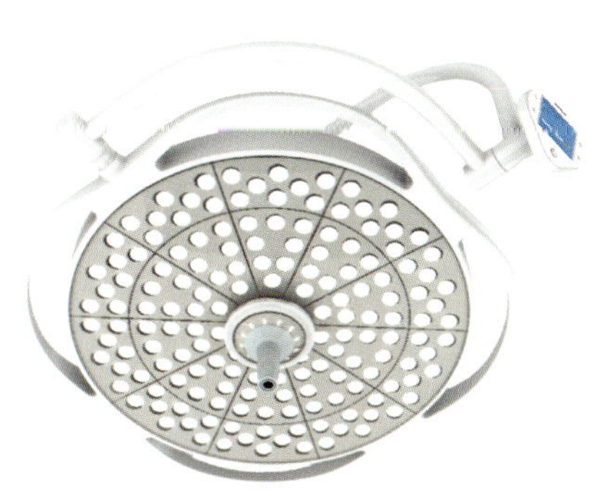

北京谊安医疗系统股份有限公司

BEIJING AEONMED CO., LTD.

无影灯LED OL9570/50

LED Medical Light Purelit OL9570/50

此款无影灯以LED技术为基础，实现了承重壳体与把手的一次成型，使得灯体更加稳固，操作的舒适性和有效性大大增强，同时也使得灯体在造型上更具有整体性；该产品新增摄像技术，能全程监控和记录手术过程，大大提升手术的质量。

Based on the LED technology, this shadowless lamp has made a great progress both in technology and design. Firstly, the integrated shell connects the light shell and handle together, which is helpful to improve the quality of using experience and makes the product look more integrated. Secondly, the newly added camera will improve the operation quality without doubt.

成都意町工业产品设计有限公司

3in1数字无线话筒

该产品为第四代无线话筒，采用2.4G无线传输技术，并具有激光教鞭、课件演示无线控制功能；产品内部零件排布合理，空间利用率极高；机身银色亮光外壳使产品看起来更加时尚高档，话筒既可以握在手中使用，也可以挂在胸前。

Chengdu Iding Industrial Design Co., Ltd.

3in1 Digital Wireless Microphone

It is the 4th generation wireless micro-phone adopting 2.4G wireless transmission technology. With a laser pointer pen and a courseware presenter, it is a customized product for meeting and teaching; the arrangement of internal parts is rational with very high utilization rate; shining silver outer cover seems more fashionable and of top quality.

成都意町工业产品设计有限公司

四方儿童游戏手柄

这是一款专为6到9岁的儿童设计的游戏手柄，小巧圆润的外壳能够让儿童轻松拿握；将功能键与方向键设计为向两侧倾斜，且均安置在拇指的活动范围内。产品表面磨砂处理增加摩擦力，避免长时间使用因手汗而滑落。

Chengdu Iding Industrial Design Co., Ltd.

SFITC Child's Game Pad

This game pad is specially designed for 6~9-year-old children. With a round and small figure, it can be easily held by children. Also the function keys and direction keys are designed to be listed to the sides and within the activity range area of the thumbs. The sanded surface increases the friction to avoid the pad slip-out because of hand sweat.

成都意町工业产品设计有限公司

骨传导耳塞

该耳塞利用固体扬声器对颅骨进行振动，直接将声音传达至听觉中枢，这种方式最大的优点是即使长时间使用也不会对耳膜造成任何损害。耳塞的把柄从下贯穿至后包裹住尾部，呈现一个独特的中国云纹图案。镀铬的把柄和黑色亚光的机体对比强烈、时尚。

Chengdu Iding Industrial Design Co., Ltd.

Bone Conduction Ear Plug

The bone conduction method has changed the traditional way of conducting the sound to the eardrum, utilizing the solid speaker to convey the vibration on the skull and transmit the sound to the auditory centre. The great advantage of this ear plug is that it will not cause any harm to the eardrum even if you use it for a long time.

成都意町工业产品设计有限公司

Chengdu Iding Industrial Design Co., Ltd.

埃森普特半自动弧焊机

EXPERT Semi-automatic Arc Welder

设计灵感来源于身躯强壮的越野车造型，张扬的外观爆发出活力，饱满有力的前后壳使用了同一套模具，同时通过改变中段钣金机箱的尺寸，占据了前脸一半面积的栅格保证了良好的散热性能；前倾的把手符合人机工学，让使用者更加省力和舒适。

The design inspiration is from the shape of off–road vehicles with strong frame, while the wild and bold appearance also goes off into its vigor. The strong front and back shells use the same modules, and through changing size of the middle framework, 10 models are produced; the grid which occupies half of the front shell has a good heat dissipation performance.

成都意町工业产品设计有限公司

Chengdu Iding Industrial Design Co., Ltd.

长江电动喷涂机

Changjiang Electric Airless Paint Sprayer

此款产品的特点即为"小巧"，作为一款便携式电动喷涂机，它的外观给使用者的第一感觉就是轻便，易用。型号与产品的PVC贴纸正好贴于凹陷处，图案和颜色给产品注入活力。把手的位置的设定于整机的重心之上，操作更轻便。

The key feature of this product is portable. As a portable electric airless paint sprayer, the user's first impression is the lightweight, portability and easiness to use. Both the model number and the product PVC sticker are pasted on the recess part, together with the figures and colors, injecting vitality to the product. With the external plastic shall demoulded in the gravity center of the whole body, it is more convenient to operate.

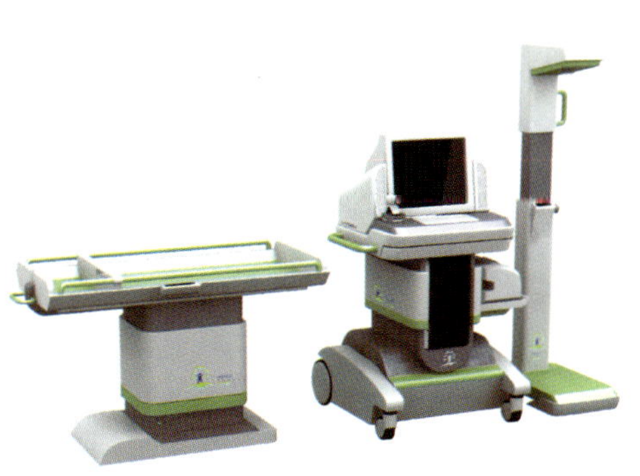

创意工场（北京）工业设计有限公司

IdeaWorks (Beijing) Industrial Design Co., Ltd.

儿童综合素质测试仪

Children Overall Quality Tester

PPVT测试具有声报图片词汇和语音自动提示性能。可在走动的字幕上按照选定的符号用手指进行点击划销或者在选中的图形上点击，也可以用鼠标进行点击。

PPVT test has the function of reporting pictured vocabulary with voice and auto–matic speech reminding. In accordance with the selected signs, one can click on the rolling caption with his own fingers or click on the selected graphics; this can also be carried out with the mouse click.

递加（北京）设计顾问有限公司

银河新星无创血糖检测仪

银河新星无创血糖监测仪采用独创的机体代谢率测量探头，检测得到的代谢率变化特征导出血糖值；采用与主流技术类似的近红外漫反射光谱分析方法，获得血糖值。产品设计上形体饱满、圆滑，突破传统医疗产品呆板简单的印象，造型曲线配合人体手掌握合曲面，便于操作。

Design Plus (Beijing) Consulting Co., Ltd.

Non-Trauma Blood Sugar Detection

Non–Trauma blood sugar detection uses an original near–infrared diffuse reflectance spectroscopy method which is similar to the mainstream technology of obtaining blood sugar level. The product's body design is full and smooth, which is a breakthrough compared with the simple, wooden impression of traditional medical products. The curved model shape coordinates with human palm perfectly, thus convenient to operate.

广东华南工业设计院

Podium USB扩充器

也许您会为难以辨识的读卡器感到烦恼，Podium将解决以上问题。它帮您扩充了三个USB接口用于连接U盘或无线设备接收器，同时配备时下最流行的四种相机、手机、游戏机储存卡直读接口，使桌面变得干净、整洁。概念来源于花盆，简洁的造型搭配突出的品牌标识，构成富有亲和力的品牌形象。

GUANGDONG SOUTH CHINA INSTITUTE OF INDUSTRIAL DESIGN

Podium

Maybe you are annoyed by the rough–and–tumble USB lines and many card readers with unrecognizable functions scattering around your computer. Podium will help you solve this problem easily by extending 3 USB interfaces, which connect USB flash disks and equipt with direct interfaces for four types of cameras, cell phones and game machine memory card. Inspired by flower pots, its concise modeling highlights the brand logo and sets up a brand image of rich affinity.

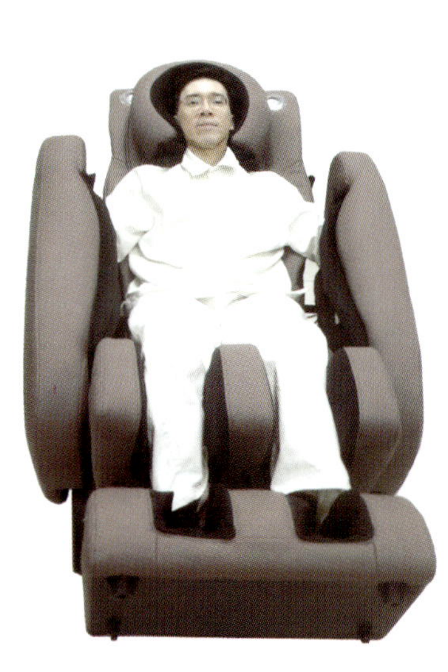

东莞市生命动力按摩器材有限公司

LP7000航天按摩椅

本产品多功能的开合机关，令手部按摩扶手开合200mm，专利人体工学设计，适合身材不同的用户。椅背上的环绕立体声音响，避免耳机对耳朵造成压力，且可连接MP3、MP4、CD及游戏机等，将音乐节奏融入到按摩中。

Life Power Health Ltd.

Massage Chair

The design was inspired by the streamline feature of aerospace craft. Provide 200mm clearance motor drive adjustment, creates best ergonomics arm massage for large and small sizes users. The stereo speakers mounted both sides of the headrest provide high quality music while massaging. The sound cable can be connected to MP3, MP4, CD and handheld game players.

多达创新（北京）科技有限公司

电脑一体机

电脑一体机可以根据性能需要随意组合，隐藏式走线设计，让显示屏通过支架内的隐藏式走线槽连接到电脑主机，采用LED背光屏，无闪烁、色彩纯正，整机耗能40W，比传统PC节能70%。

Dooda (Beijing) Industrial Design Co., Ltd.

Computer

The integrated computer machine can be combined at will according to the performance. The hidden wire scheme connects the display and main engine with hidden wire in the trestle. The adoption of LED illuminant screen, with zero twinkle, pure color and an energy consumption of 40W, can save 70% energy than traditional PC.

多达创新（北京）科技有限公司

外置吸入式光驱

随着内部光盘的运转，外置吸入式光驱的蓝色由浅至深的变化，使产品更加立体化；灰色区域略低于黑色区域，形成了一个小小的台阶，可将要播放的光盘放置在灰色区域上；在左侧面设计了一个切面凹槽，使使用者可以轻松地将吸入式光驱拿起放下。

Dooda (Beijing) Industrial Design Co., Ltd.

The Instruction of External CD Drive

This slot–loading drive form is simple, generous, practical and focusing on its human characteristics. There is a cross–section groove design in the left side of the cover so that users can easily pick the CD driver up or lay it down.

多达创新（北京）科技有限公司

电源适配器

该笔记本电源适配器的设计根据电路布局设计，合理利用空间，最大的亮点在于解决了线材收纳问题。能将线材很有条理地收纳在产品本身的收纳盒内，便于整理与携带。

Dooda (Beijing) Industrial Design Co., Ltd.

Instruction of Power Adapter

The Power Adapter Design takes full advantage of space rationally according to the circuit layout. The biggest highlight is the solution to the problem of wire collection. It can orderly collect the wire rod into the collection box, thus easy to organize and carry.

方太柏厨设计中心

雍容橱柜

“雍容”能让人感受到开启门板的轻松快捷，关闭顺畅静音。充分利用吊柜与地柜之间的空间距离，使物品的摆放或取出更加便捷，追求产品的整体造型和谐与统一；后台面功能盒为您的调味瓶、碗碟、抹布等提供合适的储纳场所。

Borcci Design Center

Polyline

"Polyline" gives people the enjoyment of relaxation and easiness while opening the door and closing smoothly and silently. It takes full advantage of the space distance between the wall cupboard and floor cabinet, making objects placing and removing more convenient and reflecting its pursuit on overall harmony and unity of the products' overall shape. The rear table's function box provides a suitable storing place for your seasoning bottles, dishes, cloth and so on.

福州瑞达电子有限公司

LED数字钟

LED数字钟，设计师将60个小方格替代60秒，随着小方格由少积多，达到60个时间格就增加一分钟，此设计将呆板的数字时间趣味化。

Fuzhou Reida Electronic Co., Ltd.

LED Digital Clock

For LED Digital Clock, designers replace 60 seconds with 60 small squares. When the quantity of small squares accelerates up to 60, 1 minute increases accordingly. This design makes the boring digital time more interesting.

福州瑞达电子有限公司

旋转镜子数显钟

旋转镜子数显钟，屏幕同屏显示时间、温度、月份、日期、星期，反转圆形部分可作镜子使用。

Fuzhou Reida Electronic Co., Ltd.

Vortical-mirror Digital Clock

Vertical–mirror digital clock shows time, temperature, month, date and week on the same screen. It can be used as a mirror when reversed.

福州宜美电子有限公司

EM1340 “田”字型投影钟

“田”字型投影钟可以在“田”字型LCD显示屏上分别显示时间、日期、星期与室内温度等内容，还可以通过180°旋转把时间直接投影到墙上，晚间无需开灯或看手机，只需朝室内任意平面照射，即可知道当时的时间，一目了然。

FUZHOU EMAX ELECTRONIC Co., Ltd.

EM1340 “田” Project Clock

It can display such content as time, date, week and indoor temperature, which can be seen in the "田" shape LCD screen. You can even rotate 180 degrees and project the time directly on the wall. In order to know the time, you just need to project the time to any plane indoors without turning on the lights or the cell phone.

福州宜美电子有限公司

EM711可旋转式太阳能温度计

可旋转式太阳能温度计的背面安装太阳能片，采用太阳能供电或内置锂电池辅助供电，不仅经济环保，而且采光度强，适合室内、室外摆设。头部为活动设计，可进行360°旋转，在任意角度都可以读取数据。

FUZHOU EMAX ELECTRONIC Co., Ltd.

EM711 Rotatable Solar Thermometer

The back of the rotatable solar thermo-meter is installed with a solar sheet, its power is supplied by solar or lithium battery. This is not only economical and environmental but also carries strong light function, making it suitable for indoor and outdoor decoration. The head part is especially designed for movement and can rotate within a range of 360 degrees, making it possible to read data from any angle.

高国兴

互拷器A2B

当周围没有电脑或笔记本的情况下，该产品能够满足我们日常方便需求。产品有两个USB接口，下面称之为A，B。产品小巧便携能通过产品的液晶屏幕读取A，B电子数据资料，并且能够通过本产品实现A，B资料互拷。

Guoxing Gao

A2B MP3

When a computer or laptop is not available for the moment, the product can meet our daily needs for convenience. The product has two USB interfaces, which are used to link USB disks or other equipments with output digital products. Below they will be referred to as A and B. As a compact and portable product, it not only supplies us with electronic data of interfaces A and B from the LCD screen, but also realizes mutual data copy between A and B.

光彩无限创意中心

老年人服务平台终端

老年人服务平台终端增加了针对老年人需求的功能，如：移动定位、后台指路导航、摔倒自动报警、计步功能、语音报读功能、服药提醒功能。简洁的外观，简单的拆装模式，醒目的按键及超大的屏幕显示，都传达对老年人的无限关爱。

Shine Design Consultation Center

Digital Service Terminal for the Elderly

Digital Terminal for senior citizens, which is completely designed for the age group, carrying many practical functions. It caters for senior citizens' demands especially, such as: mobile positioning, background navigation, automatic fall alarm, step counting, and voice call, voice function, one-touch service, medicine reminder function and so on. Concise appearance, simple disassembly mode, eye-catching buttons and large screen display all together convey the infinite care for the elderly.

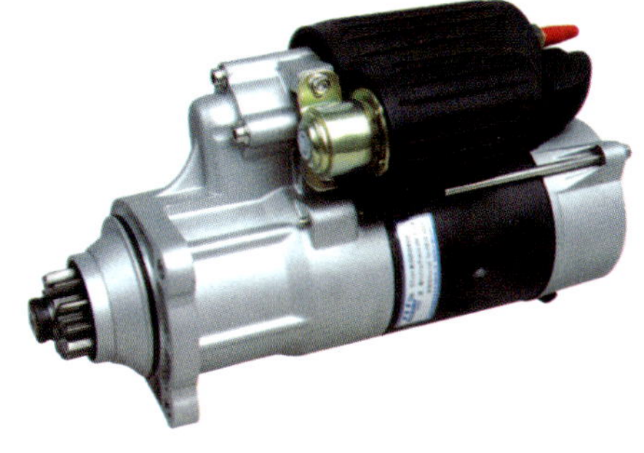

光彩无限创意中心

起动机M105R3040SE

起动机M105R3040SE通过设计降低了原有机械制造的冰冷感，将力与美相融合，展现出"新"的魅力。该起动机带着黑的稳重、银的灿烂，一经推出，迅速得到市场的认可，占据80%的市场份额。

Shine Design Consultation Center

Starter Motor M105R3040SE

The new design of Starter Motor M105R3040SE has taken full account of the strength and the structure requirements, at the meantime highlights the concept of creativity in every detail by combining strength and beauty. Once has been launched, it is recognized by the market and occupies 80% of the market quota.

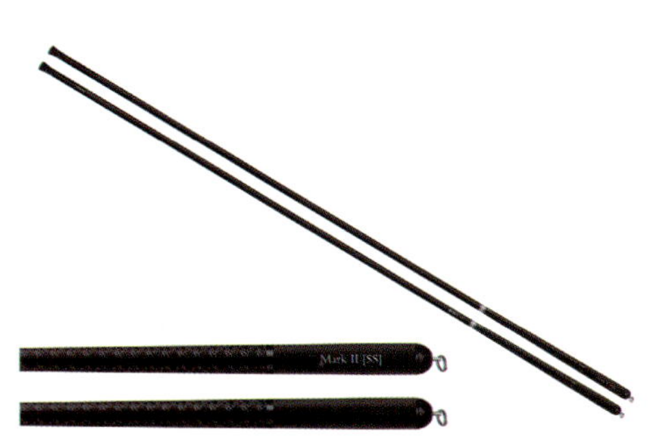

光彩无限创意中心

鱼竿—碧波2S

方形设计避免了鱼竿间摩擦碰撞导致的表面喷涂脱落或钓竿滚入水中的情况。"定堵"上的透气孔，利于平日收藏时主竿内部支竿的干燥、透气。手把处的防滑花纹及"定堵"上鱼形透气孔，在满足功能及美观性的同时，更具备品牌的识别性。

Shine Design Consultation Center

Fishing Rod-BIBO 2s

The plug1 of fishing rod –"Rolling stop" is designed into a square shape to avoid friction between fishing rods and falling–off of the rods. The holes on "Rolling stop" are conducive to volatilization of water vapor, ensuring successful entry of the dry air into the fishing rod. The antiskid figures on the handle and ventilation holes not only satisfy the function and aesthetic property, but also make the brand more recognizable.

广东美的环境电器事业部

GD Midea Enviroment Appliances MFG. Co., LTD.

加湿器S30U-H

Humidifier S30U-H

该设计将加湿器改为可爱、情趣化的、家居装饰性强的实用型家居电器。采用了斜向上的出雾方式，避免了垂直出雾时，水雾落到桌面弄湿家具。在空调房、秋冬等干燥的季节，给身体和皮肤补充水分，滋养肌肤，同时预防哮喘等吸呼道疾病。

This is a kind of decorative humidifier. In air-conditioned rooms and during dry autumn and winter, it adds water to the body and nourishes the skin, thus preventing asthma and other respiratory tract disease.

广东时尚元素生活用品有限公司

Guangdong SSYS Living Goods Co., Ltd.

DIY积木概念灯

DIY Block Concept Light

DIY积木概念灯选用环保、耐热的材料，通透的模块组合构成。灯体有模块形成的空格，是为了有更好的散热效果。积木概念灯模块有红色与蓝色的区分，用蓝色的模块接近灯源，为正常照明的效果；用红色的模块接进灯源，为浪漫气氛的效果。

DIY block concept light is made from green and heat-resistant material, compromising of penetrating module blocks. The blank space between the blocks is to create better heat loosing effect. As for color, Red and blue are adopted as perfect combination: Putting the blue modules around the light we can get normal lighting, and putting the red modules around the light is for romantic ambience.

广东新宝电器股份有限公司

GuangDong Xinbao Electrical Appliances Holdings Co., Ltd.

即热式水壶

Hot Cup

即热式水壶产品健康、节能，它采用高效即热专利技术，26秒即出开水。同时采用一键启动，自动出水，烧水出水一气呵成，快捷健康安全，无需等待。整体造型时尚，水箱内置灯光绚丽且便于观察水位。

Hot Cup has a high efficiency heating system, which makes the water boil in 26 seconds. It also features one-button startup and automatic water outlet, thus safe, healthy and no need to wait.

广东新宝电器股份有限公司

多功能早餐机

多功能早餐机集水壶、多士炉、咖啡机的功能于一身，方便快捷、节省空间，具备简约的外型和时尚的外观。

GuangDong Xinbao Electrical Appliances Holdings Co., Ltd.

Multifunctional Breakfast Machine

Multifunctional Breakfast Machine integrates the functions of kettle, toaster, and coffee maker all together, thus it is very convenient, fast and space-saving. Its appearance is also very concise and fashionable.

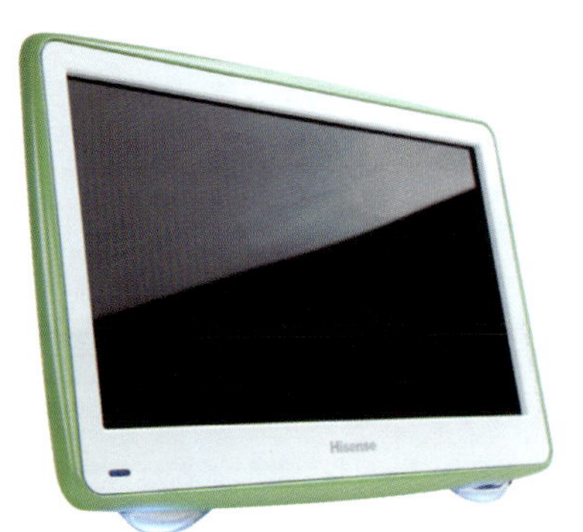

海信集团

V09系列儿童互动电视

这是海信设计的首款专为儿童开发的平板电视产品，采用触摸屏方式操作，内置学习和娱乐软件。外形设计以体现童趣为核心，卡通风格的造型和色彩搭配能吸引儿童的注意力。

Hisense Group Limited Company

Hisense V09 LCD TV

The V09 is Hisense's first portable, multi-functional television designed especially for children. A touch-screen improves user's experience. The unit is packaged with games and educational software such as language learning programs, Internet browsing ability, chess, and cartoons. Exterior design takes it as a core work to reflect the playful element, while the cartoon-style shape and color matching can attract children's attention.

杭州德意厨具有限公司

自由舰橱柜

“自由舰”融合了建筑与家装设计风格，凸显产品的时尚元素；依据人机工程学原理，将台面分成两个台阶，全新的线性分割设计，更加睿智。色彩上着重将北欧设计风格引入国内橱柜文化。

Dandy Holding Group Co. Ltd.

Free Ship

Integrating architecture design and decoration, "Free ship" highlights the fashionable products; the overall shape brings bold change into surface design; basing on ergonomics, new linear segmentation is adopted, the table is more wisely divided into two steps. As for color, it focuses on the introduction of Scandinavian design culture into domestic cabinet.

杭州老板电器股份有限公司

HANGZHOU ROBAM APPLIANCES Co., Ltd.

侧吸式油烟机5360

Icook5360

5360清爽、干净的外观是最大亮点。方形的机身，利于工艺量产和厨房搭配；圆形的导流板既可遮挡油网、油路等复杂的机械化结构，又能引导风向，增加吸风压力，还能充当进风区挂油收集的最佳托盘。

Icook 5360 features a brisk and clean exterior appearance. Circular flow-guiding board not only firmly keeps out complicated mechanization structure but also can guide the wind direction, increasing the suction pressure and acting as the best oil tray linked to entering wind area.

杭州老板电器股份有限公司

HANGZHOU ROBAM APPLIANCES Co., Ltd.

聪明套装

Smart08

聪明套装旨在建设智能、洁净、轻松的厨房。在自动模式下，油烟机吸风量将根据燃气灶火焰大小自动调节。燃气灶采用触摸控制，方便用户清除油污。独立双模消毒柜互相独立隔离，满足不同的使用需求。

This smart package is designed for an intelligent and relaxing kitchen environment. This range hood, which combines with frequency conversion electric motor, saves energy and protects the environment. Under automation pattern, the range hood adjusts the air suction rate automatically in accordance with the gas cooking stove's flame size, giving convenience to grease dirt elimination.

浩汉工业产品设计（上海）有限公司

Nova Design Co., Ltd.

美的凡帝罗三门冰箱

Midea "Vandelo" 3-Door Refrigerator

美的凡帝罗三门冰箱，L型把手结合了视觉语意和功能使用上的考虑，给用户留下深刻印象同时操作更加人性化。极简的外观设计，体现了专业美学并且能够轻易地融合在时尚的家居环境里。

Midea Vandelo Refrigerator pops its characteristic with a very iconic "L" shape handle, combining image and function perfectly. This handle humanizes overall operation. Minimalist design reflects the professional aesthetics and can be easily integrated into the stylish home environment.

鹤山市洁丽实业有限公司

HESHAN CITY JIELI ENTERPRISES Co., Ltd.

长虹饮涧JL7913500水龙头

Arch Bridge JL7913500

这款作品在造型上采用简约的弧线，巧妙勾勒出独特的外观和丰富的桥梁神韵。在使用角度和出水角度方面也充分考虑了共架沟通、共处和谐，这正是设计师想表达的寓意。

Concise arcs are adopted in the design of this product, forming a special appearance and infusing verve of an arch bridge. Through sufficient consideration of both the using angle and water flowing angle, the designer's intention to show an understanding and harmony is embodied completely.

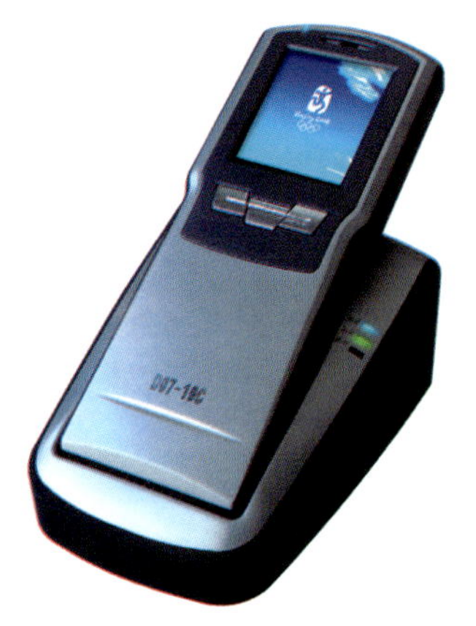

恒创碧思特工业设计（北京）有限公司

BEST IDEA INDUSTRIAL DESIGN Co., Ltd.

消防安全管理及监督系统信息终端

Fire Safety Management and Monitoring System Information Terminal

消防安全管理及监督系统信息终端通过采用RFID、数字图像处理、网络、通信相结合的即时信息传递技术，将各单位法定安全责任的履行情况即时提供给相关责任人，同时记录检查的影像，以保证各项消防安全管理工作的落实。

By using the combined instant messaging technology of RFID, digital image processing, internet and communication, fire safety management and monitoring system information terminal delivers real-time security implementation conditions to the relevant responsible person.

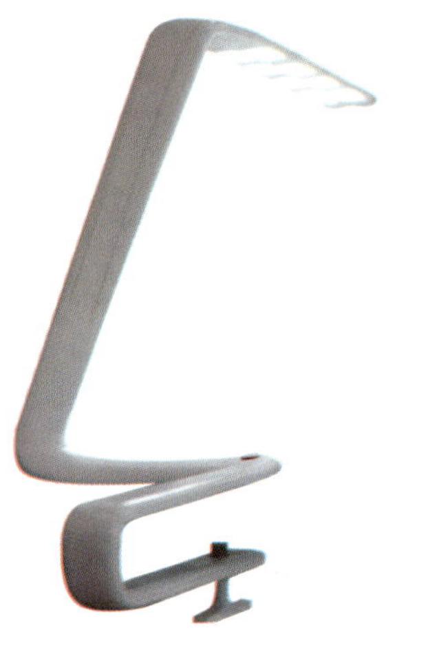

黄治中

Zion Huang

LED工作台灯

Stretch

本设计利用LED光源体积小，耗能低，能大尺度摆列的优点，形成一个线性延伸的光线覆盖工作台面，使每个区域都有均匀的光线，使用者在任何位置都能享受相同强度的照明。

By utilizing the advantages of LED light-small size, low power consumption, this product forms a linear extended work table covered by light. Thus it ensures that all areas of the table share uniform light and users in any location can enjoy the same intensity of illumination.

桔思创意设计

香料娃娃

香料娃娃将调味料先倒入开口的勺子中，看看分量多少，再根据用户的口味添加到菜品中，方便控制调料分量保护健康。

U.I.Design Co., Ltd.

TeruTeru Spice Doll

From the sorting spoon at the top of TERUTERU, you will easily see the amount of seasoning and then adjust to your own taste, so you can comfortably enjoy both a beautiful meal and a healthy life.

桔思创意设计

沙拉碗

针对用户量身定做的沙拉碗，马丁尼式的碗身设计，单手即可拿取，随处可用；弧形叉头设计，使用方便又有趣。

U.I.Design Co., Ltd.

Salut

Salut is a special customized salad bowl looking like a martini cocktail glass. It can be easily carried around with a single hand in a party. The specially-designed fork can pick any shape of vegetables and matches the bowl perfectly.

桔思创意设计

点心碗

针对用户量身定做的点心碗，马丁尼式的碗身设计，单手即可拿取，随处可用；加上特别的不锈钢勺，简约流线型设计，使用方便又有趣。

U.I.Design Co., Ltd.

Salut BonBon

SalutBonBon can be easily carried around with a single hand in a party. The specially-designed stainless steel spoon is simple and stylish, making a perfect fit with the bowl.

康佳集团

V12手机

整机采用不同深浅的枪色电镀，低调的光泽尽显高贵和品质。采用windows mobile操作系统和3寸纯平触摸屏。拥有wifi和GPS功能。触摸式导航键，操作方式类似ipod，简单方便。

KONKA GROUP

V12 Mobile Telephone

The machine uses the different depth of gun color for electroplating, while the twinkling gloss manifests nobleness and quality. It is a smart phone with windows mobile os and 3.0" touch-screen. The handset has wifi and GPS functions. The navigation key can be touching controlled, owning an operating mode similar to iPod.

康佳集团

9900C手机

KONKA 9900C 是一款专门为老年人和盲人设计的手机，低端直板设计，单色显示屏，大按键大字体设计，具有通话免提、手电筒、语音报时、数字按键报读、SOS紧急呼叫及快速拨号功能。

KONKA GROUP

9900C Mobile Telephone

KONKA 9900C is especially designed for the old and the blind. Low-end bar design, mono-color display and super big key and big fonts design. It carries the following features: perfect hands-free talking, electric torch, voice time teller, digit teller, SOS emergency call and shortcut dialing.

康佳集团股份有限公司

变频节能星81系列

变频节能星81系列液晶电视采用的AGT节能屏、OPC节能芯与PMS节能系统，大大提高了其产品的节能性能，系统节能可达52%。选用无痕注塑面板及凹陷式红色回音槽。机身巧妙运用光学原理，红黑渐变自然平滑，诱惑动人。

KONKA GROUP Co., Ltd.

Konka Frequency Conversion Energy-saving Star Series LCD TV

Frequency-Conversion Energy-saving Star series LCD TV adopts AGT energy-saving screen, OPC energy-saving core and PMS energy-saving system, enhancing its energy-saving performance up to 52%. The innovative dented red echo slot on the E-mold panel makes the sound wave reflect well.

康佳集团生活电器

康佳电压力煲

本产品整体造型流畅连贯，顶盖与面板通过酒红色镶件有机连成一体；机身棱线分明，更具立体感，大显示大按键的设计让产品更加人性化，使用方便。

KONKA GROUP Co., Ltd.

Electric Pressure Cooker

Modern popular elements are adopted in the design of this product. The top cover and the main body are well combined together by a claret bar. Big display screen and buttons offer users a better operation experience.

可能贸易（上海）有限公司

多功能水壶

携带净水壶由净水壶体、过滤芯、自动计数器、内胆四部分组成，注入水并盖好盖子后，计数器能自动倒计数，显示屏若显示出过滤芯还可使用的次数到0，就表示使用者该换过滤芯了。

KANON TRADE (SHANGHAI) Co., Ltd.

Portable Purifying Kettle

Portable purifying kettle is comprised of water bottle, filter, automatic counter and liner. When the counter starts time counting automatically, the display screen would show the remained usable times. When it shows "zero time", the filter needs to be changed.

李宁（中国）体育用品有限公司

化石渗水鞋

李宁化石涉水鞋的侧面设计如同肋骨，为足部侧向运动带来了稳定性。鞋头部分的橡胶避免脚趾被水中的砺石碰伤。鞋底的设计犹如三叶虫的结构，能有效防滑。整双鞋子与拖鞋大小无异，可作为备鞋携带。

Li Ning Co., Ltd.

Fossil Aqua Shoes

A better solution of aqua shoe. People use it as a tool to walk through rill. The shoe's upper structure which seems like rib brings plenty of lateral support. The pattern of the sole is inspired by trilobite with two directions from toe to heel. Fossil has a small volume as slipper, thus can be taken as reserved shoes.

李宁（中国）体育用品有限公司

猫爪五代野外跑鞋

李宁猫爪五代野外跑鞋，搭载T–Cushion减震技术、ProBarLOC足弓支撑技术及防刺穿板、侧向防护等科技，此外创新运用内嵌式橡胶页片防翻转系统，双V字防压迫束紧系统等新技术，全曲线底部颗粒更适合野外地面状况，提供稳定的抓地性能。

Li Ning Co., Ltd.

X-claw V Trail Running Shoes

Li Ning X–claw V Trail Running Shoes have many features: being equipped with T–Cushion shock absorption technology, ProBarLOC arch supporting technology and anti–piercing plate, lateral protection and so on. Innovative use of many new technologies, such as anti–rollover system, double anti–oppression V–tie system, the whole curve of the bottom is more suitable for field conditions and provides a steady grip on the ground.

李宁（中国）体育用品有限公司

G–Shark篮球鞋2

这是一款为顶级专业比赛而打造的篮球鞋，独特的鞋身造型灵感源自大鲨鱼，其中模拟鱼鳃结构的VENti–gill换气系统，有效解决了传统篮球鞋透气性差的问题。鞋面选用环保天然磨砂皮以增加包裹性能和穿着舒适度。

Li Ning Co., Ltd.

G-SHARK II Basketball Shoes

G–Shark is a special design for top professional basketball players. The unique reversed–arrangement lateral and medial structure can effectively improve the air ventilation condition of the shoe, which solved the problem of inferior heat elimination of traditional basketball shoes. Leather uppers adopt environmentally–friendly natural nubuck in order to increase package performance and wearing comfort.

联想（北京）有限公司

IdeaPad Y650 联想

Ideapad Y650是全球超轻超薄的16英寸笔记本产品，采用16:9黄金比例LED背光液晶屏。Family ID Y650拥有全新的时尚外观设计，神秘的棕黑色中透射出经典的六边形图案组合，靓丽的白色C面、发光的ideapad logo闪动您的视觉。

Lenovo(Beijing) Ltd.

IdeaPad Y650

IdeaPad Y650 is the world's thinnest 16–inch entertainment notebook. The Y650 is extremely thin and light for a 16–inch notebook. At a 5.6–pound–level, it's lighter than many 15–inch notebooks while its performance is on par with some 14–inch notebooks. It has a dark textured lid with a geometric pattern that resists fingerprints, white interior and orange trim.

联想（北京）有限公司

IdeaCentre A600

全世界最薄的一体电脑Idea Centre A600外观设计灵感源于"芭蕾"造型。仅一根连线，让你从此告别桌上电脑带来的凌乱感。"中国风"设计元素的应用及设计于侧身的常用接口，让一切变得简约典雅。

Lenovo(Beijing) Ltd.

IdeaCentre A600

The Idea Centre A600's all-in-one desktop system combines the function of a traditional tower PC with the style and simplicity of a television. Powerful computing, easy installation, space-saving design and high-definition multimedia are all contained in one very slim and attractive package. The application of "China chic" design elements and common interfaces on sideways makes everything become simple and elegant.

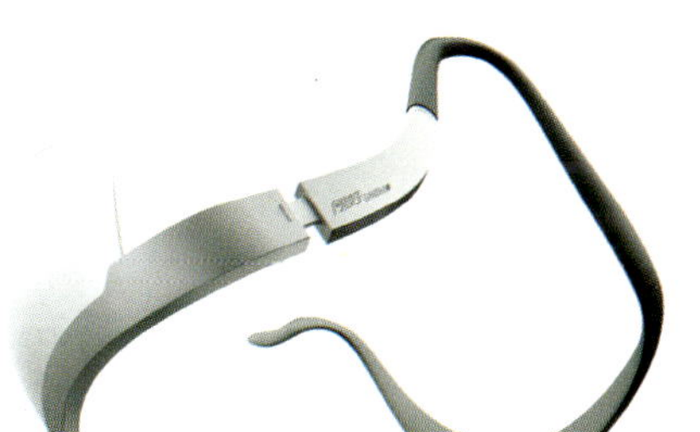

洛可可工业设计公司

防唾液挡

防唾液挡专为餐饮行业人员设计。外观上既自然，又富有清新自然的生活气息；结构安排完美，它可以有效地阻挡住工作者在工作过程当中口腔的分泌物对食品和周围环境造成的污染。

LKK Industrial Design Co., Ltd.

Impediment Saliva Card

This product is specially designed for people in the catering industry; it can effectively prevent saliva pollution on food and nearby environment in their working process. The product embodies a natural fresh and modern design, and it is well organized in structure.

洛可可工业设计公司

"竹·玉"系列水杯

"竹·玉"系列保健杯不仅将健康优雅的生活品质表述无疑，更是浸透出东方人的典型文化取向，一"竹"，一"玉"，外孕其水，内养其里，相得益彰。同时，采用绿色环保材料，反映出"水宜生"的品牌理念。

LKK Industrial Design Co., Ltd.

Alkaline Ionized Water Generator

Alkaline Ionized Water Generator embodies healthy lifestyle concept and integrates Chinese people's psychology: to pursue perfect appearance as well as inner good quality. Environment-friendly materials indicate the brand concept of "water is good for life".

洛可可工业设计公司

LKK Industrial Design Co., Ltd.

骑士安全帽

Knight's Helmet

该产品采用了进口ABS材料；顶部两侧设计了侧向透气孔，避免进水；模具标识采用了反光漆喷涂工艺，在暗中遇直射光线便更加明显。另外，帽尾内侧设计的透明卡式署名槽，可有效避免混戴问题。

The brand-new shell uses imported ABS material. Air holes designed on the lateral top instead of straight top can avoid water inflow. The logo adopts reflector paint technology, reflecting the light more obviously. In addition, in the rear of a helmet is a transparent cassette sign preventing people wearing others' helmet by mistake.

洛可可工业设计公司

LKK Industrial Design Co., Ltd.

梓路苑路灯

Leading lamp for "Ziluyuan"

"方正之气"采用现代设计手法，将矩形延展成错落之方。黑与金的碰撞，表露出高贵肃穆的汉代奢华。漫漫寂夜之中，它正直挺立，既融于梓路苑幽静、神秘、肃穆的整体氛围，也表达出对逝者的诚心敬重之情。

This product adopts squares of different layers. Black and gold colors embody dignity and luxury. It stands erect at night, not only fits the mysterious and quiet atmosphere, but also shows respect to the deceased.

洛可可工业设计公司

LKK Industrial Design Co., Ltd.

黄山度假村路灯—"徽屋"

Street Lamp For "Qishu"

这款道路灯的设计灵感来源于徽派建筑的神韵。设计师将"粉壁、青瓦、马头墙"的徽屋特点揉合进了灯具中。聚光主照明灯和散光辅照明灯两个光源，使得灯体在满足道路照明功能的同时，传达出温暖的家的讯息。

The shape of this street light is inspired by the architecture in Anhui province, especially the characteristics of "pink walls, grey tiles and corbie steps".One spotlight and one sidelight give out enough light on the road, and at the same time conveys the signal and warmth of home.

佛山市顺德区美的电热电器制造有限公司

美的电磁灶CT2101

该电磁炉可完全实现台嵌两用，加热区域的炫蓝火力圈显示解决了以往产品火力不可见无法掌控火力的问题；语音提示能防止用户误操作；滑动式触摸火力控制便于用户调控火力；铝质镂空外框便于空气流通。

FoShan ShunDe Midea Electrical Heating Appliances Manufacturing Company Limited

Induction Cooker Midea CT2101

This induction cooker can be used both embedded in the wall or on the cooking bench, the shining blue fire demonstration and sliding touch-control allow users to control fire easily; voice prompts can avoid users' mistaken operation.

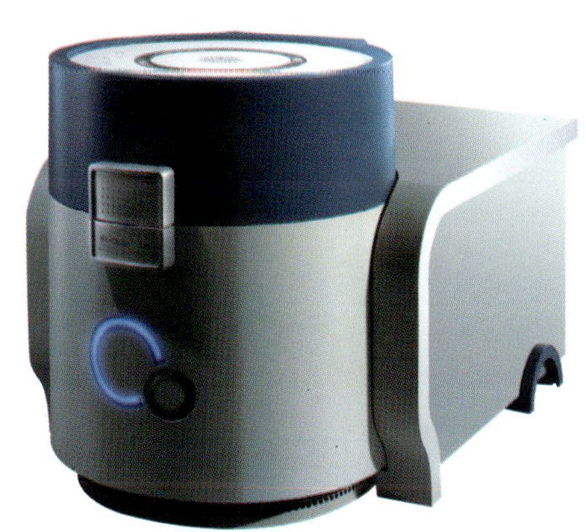

清华大学美术学院工业设计系

晶芯实时荧光定量PCR仪

这是生命科学研究和分子诊断应用领域的新型产品：采用离心式实时荧光检测方式，通过热循环PCR对特异性的靶基因进行扩增并定量。适用范围广，准确性高，体积小巧，操作简便。

The Industrial Design Department at Academy of Arts and Design,Tsinghua University

CapitalBio RT-Cycler Realtime Fluorescence Quantitative PCR

As a new product applied in the field of life science research and molecular diagnostics, Real-time Fluorescence Quantitative PCR is based on centrifugation real-time fluorescence detection. With its precise temperature control, specific target genes can be amplified and quantified accurately. Its features include wide application, high accuracy, compact size and easy operation.

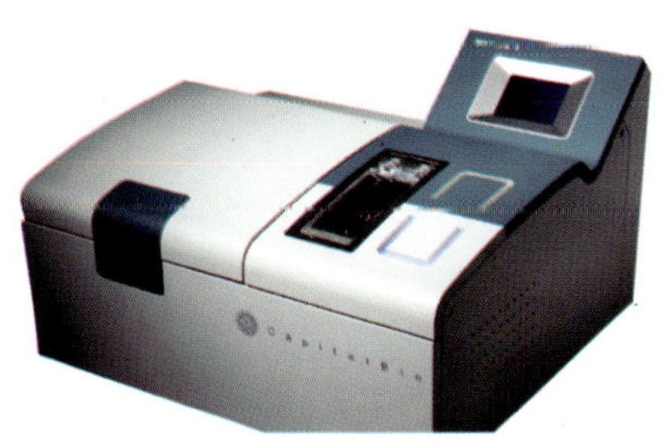

清华大学美术学院工业设计系

晶芯芯片洗干仪

晶芯SlideWasher芯片洗干仪是一款用于基因或蛋白芯片温浴清洗、离心干燥的新产品。采用实时加热、控温和离心干燥，特别设计的芯片架可重复使用，与传统手工清洗相比，提高了清洗的可靠性。

The Industrial Design Department at Academy of Arts and Design, Tsinghua University

CapitalBio SlideWasher™ 8

Slide Washer 8 is developed for redundant residual clean-up from DNA or protein microarray slide after the reaction. It integrates washing buffer heating, slide washing and slide drying functions, simplifying and enhancing the reliability of manual operations.

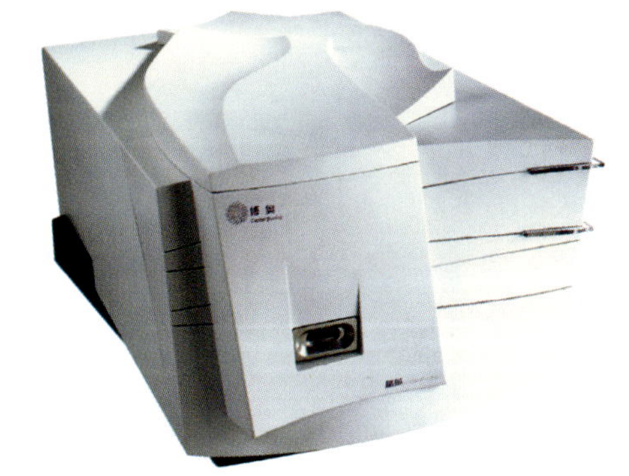

清华大学美术学院工业设计系

晶芯LuxScan双激光共焦扫描仪

晶芯LuxScan双激光共焦扫描仪采用光学、信号处理和运动控制系统等十余项技术，用于各种不同密度微阵列芯片的检测与分析，并对DNA的形态语言作了语义学层面的解构。

The Industrial Design Department at Academy of Arts and Design, Tsinghua University

CapitalBio LuxScan Microarray Scanner

CapitalBio LuxScan Microarray Scanner is a compact high performance scanner system for micro array imaging and data analysis of DNA, protein, cell and tissue arrays. It has two excitation lasers (532 nm and 635 nm) that provide high detection sensitivity, accuracy, dynamic range and linearity for selection.

清华大学美术学院工业设计系

晶芯48微阵列芯片点样系统

该产品旨在满足生命科学领域对试剂微量化、操作自动化和高速高精度的需求。操作简便，能在多种材质的培养基表面喷洒液体样本，有良好的准确性和灵活性。

The Industrial Design Department at Academy of Arts and Design, Tsinghua University

CapitalBio Smart Arrayer 48-A Dual Purpose Microarray Spotter

CapitalBio Smart Arrayer 48 is equipped with dual printing systems: contact print-head and proprietary non-contact spray head. With high precision and maximum flexibility, the multi-purpose microarray spotter is easy to use and can print liquid samples onto various substrates.

上海杰康医疗器材有限公司

杰康鞋管家JK-A-00

杰康鞋管家采用数码微电脑臭氧负离子和脉冲空气循环净化技术，能杀菌、除臭、除湿、暖鞋和除尘。造型美观、能耗低、安全环保、使用方便。

SHANGHAI JIEKANG MEDICAL INSTRUMENTS Co., Ltd.

"Jie Kang" House Keeper For Shoes

"Jie Kang" house keeper for shoes introduces digital microcomputer ozone anion and the pulse air cycle purification technology, thus it can sterilize, eliminate the bad smell, dehumidify, warm and remove dust. It is visually pretty, energy efficient and easy to use.

上海美斯凯实业有限公司

苹果风ipao系列跑步机

该产品采用感应式按键，超大LCD显示，独创塑身、美体等6大健康管理模式和模拟机场跑道安全警示灯。外接MP3接口，配有HI–FI高保真音响、振动按摩腰带及独特的音箱孔和仰卧起坐架，适合全家健身使用。

Shanghai Maxcare Industrial Limited

Ipao Treadmill

This product features tactile keys, large LCD display, and 6 health management modes such as body shaping, as well as simulating airport runway warning light. It also has mp3 port, HI–FI, vibration massage belt and a rack, thus suitable for the whole family's fitness.

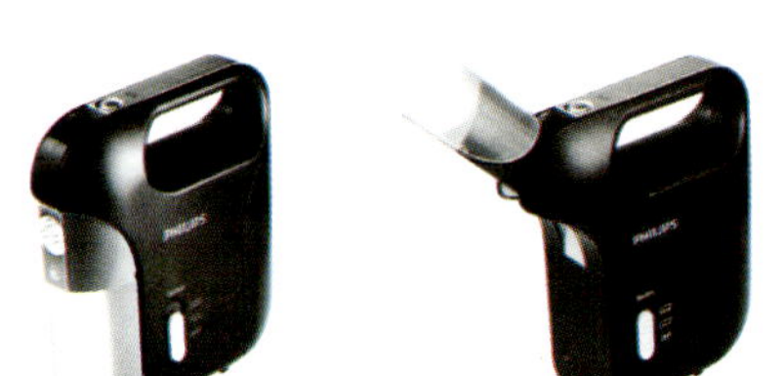

上海木马工业产品有限公司

飞利浦轻便手提灯

飞利浦轻便手提灯CZS100，由LED手电和2根节能灯管组成，作室内台灯和室外手电之用均可，能时刻满足生活所需。操作方便，内置大容量电池，经久耐用。

Shanghai Moma Industrial Product Design Co., Ltd.

Philips Portable Lantern

Philips Portable Lantern CZS100, with its easy and convenient hand–carry lamp, can be regarded as a standard lamp or an electric torch. LED switches on automatically during power failure. Dual level brightness control is easy to operate. Built–in high–capacity battery is durable and lasting.

深圳市海洋王照明技术有限公司

多功能强光巡检手电筒

该产品已获国家专利，采用双端双极无污染锂电池及可回收的航空铝合金外壳，环保性及安全性好；外形轻巧，价格低廉，适用范围广；工艺成熟，适合批量、快速组装生产；外观有防滑处理，色彩及质感给人稳重协调之感。

Shenzhen Ocean's King Lighting Science & Technology Co., Ltd.

Multifunctional Strong Light Patrol Flashlight

Multifunctional strong light patrol flashlight has acquired domestic patent, adopting lithium battery with double end, double pole and the aero metal crust resulting in good safety and environment protection. It is light, exquisite, inexpensive and applicable for various sectors. Mature productive technology fits for fast mass–production. Besides, it has experienced anti–skid treatment. The color and quality feeling give us a feeling of harmony, coordination and steadiness.

深圳市海洋王照明技术有限公司

矿灯

矿灯采用磷酸铁锂电池配以矿灯保护器，并在灯头和电池盒的结构方面加强了抗冲击的设计，提高了安全性；电池使用寿命是普通电池的两倍，和包装盒一样可以回收利用；线条优美流畅，给人耳目一新的感觉。

Shenzhen Ocean's King Lighting Science & Technology Co., Ltd.

Miner's Lamp

The protector and the strengthened impact-resistant design at the lamp holder and battery box enhance the safety. The cycle life of ferric phosphate lithium battery is twice as long as that of common lithium battery. Both battery and paper box package can be recycled and re-used. Exquisite, streamlined appearance makes people feel visually refreshing.

深圳市海洋王照明技术有限公司

遥控探照灯

遥控探照灯应用智能仿生原理，模拟章鱼设计，在30米的遥控范围内，可自由旋转弯曲，同时能强力吸附；30米无线智能遥控防震节能，高效实用；便于折叠，安全方便；价格低廉，适用范围广；材料环保，可回收利用。

Shenzhen Ocean's King Lighting Science & Technology Co., Ltd.

Remote-control Searchlight

Remote searchlight is designed on the basis of "intelligent bionic".In the light of the mollusk octopus, this product can revolve for 180° from up to down and 360° from left to right through remote control within a remote control distance of 30 meters. Its features also include the convenience to fold, reasonable price, wide application, environment friendliness and recyclable materials.

深圳市浪尖工业产品造型设计有限公司

双喜方煲・铂睿

铂睿融压力锅和电饭煲的功能于一体，排气泵设在顶部后方，便于清洗，更安全实用，并且缓解了环境空间的压力问题。在设计哲学上，本产品既带有幽默的态度，又有实用精神；都市化的风格中透露出另类却优雅的美感。

SHENZHEN ARTOP Industrial Design Co., Ltd.

Metaligen Cooker

Metaligen combines the functions of pressure cooker and electric cooker. The exhaust set on the rear of top makes it easier to dissemble when cleaning and safer when working. Moreover, it reduces the environment pressure. Concerning the design philosophy, this product combines the attitude of humor and the spirit of new classicalism, indicating modern city style as well as elegance and uniqueness.

深圳市浪尖工业产品造型设计有限公司

SHENZHEN ARTOP Industrial Design Co., Ltd.

万利达豆浆机

MALATA Blender for Soybean Milk

万利达豆浆机外观上采用女性手提袋样式，这种非圆筒形的机身缓解了打豆时需加筋扰流的困扰，有效降低了能耗和噪音。整体形态简洁、大方，富有科技感的触控按键的应用带给使用者新的生活体验。

The inspiration of the MALATA design comes from reticules that women love best. Out-of-round body can avoid operation pattern on inner surface to help the liquid flow. In this way, it saves power and reduces noise. Besides, the inductive buttons with the feeling of technology bring users a brand new experience.

深圳市浪尖工业产品造型设计有限公司

SHENZHEN ARTOP Industrial Design Co., Ltd.

多彩移动数字电视

DLA-690B

本移动电视便于用户随时随地接收电视信号，同时兼容音视频等媒体的播放，以实现娱乐的最大化。造型上将主要的按键分布于右侧，与左侧天线帽形成呼应，既增强对称美感，又凸显了简洁、大屏的视觉效果。

This product not only makes it convenient to watch TV anytime and anywhere, but also can play audio and video. The keypad placed on the right side and the antenna placed on the left together highlight the big screen, adding conciseness to the appearance.

深圳市浪尖工业产品造型设计有限公司

SHENZHEN ARTOP Industrial Design Co., Ltd.

游戏键盘T9

Gaming Keyboard T9

该游戏键盘设计中以人机工学为主线，着重考虑使用舒适度；参考了多款目前流行的和未来一年内将要发行且关注度较高的游戏，让游戏玩家参与设计，使该键盘使用更方便。外观上线条硬朗，视觉上充满力量感。

The designers of this keyboard take the ergonomics into account, focusing on comfortable operation. Moreover, they did research on popular games in nowadays and some demo games for next year and invited gamers to participate in the design, making it convenient to control. The line on the keyboard body is very hard, bringing gamers the feeling of strong power.

深圳市浪尖工业产品造型设计有限公司

SHENZHEN ARTOP Industrial Design Co., Ltd.

便捷式激光风湿治疗仪器

Portable Laser Rheumatism Therapy Device

该产品结合激光技术和传统的火针治疗，能改善局部微循环和病痛部位的营养供给。一次性塑胶套，安全检查装置，尾部急停开关，使用时安全卫生；前端的红色丝印线使选穴更快更准；外形适合手握，符合人机工程学原理。

Portable Laser Rheumatism Therapy Device combines traditional Chinese acupuncture with modern laser technology, improving blood circulation and avoiding the risks existing in point selection. The disposable rubber cover, emergency switch at the end of the device enhances safety and sanitation. Red line at the front tail makes the point selection more accurate. Ergonomic handle design makes the holding more comfortable.

深圳市企业形象设计有限公司

Shenzhen Rongyu Design & Associates Ltd.

魔力猫减压钟

MOODICARE COLOR-CHANGING

减压钟引进国外最新研究成果，运用微电脑控制研制而成。依靠12种不同的热光灯色彩的变换起到愉悦心情、缓解压力和消除疲劳的作用。

Introducing overseas latest research achievements, this lovely digital alarm clock produced by microcomputers makes you relaxed and comfortable through color change of 12 LED.

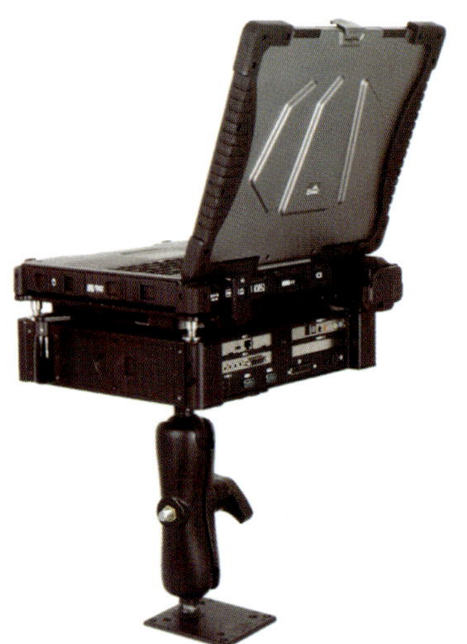

深圳市研祥通讯终端技术有限公司

Shenzhen EVOC Communication Terminal Technology Co., Ltd.

坚固笔记本

RuggedPad

该产品主要用于恶劣环境及移动的行业。网状外壳、强化橡胶包边、硅胶门设计，可有效宽温、抗振、抗压、抗跌、防水、防尘、防电磁干扰；屏幕经防反射膜处理；外观设计灵感来源于《变形金刚》，富有机械美感。

This RuggedPad is designed for industries in terrible environment and mobile works. Reticular aluminum–magnesium alloy molding, enhanced rubber covered edges and silica gel processing contribute to extended temperature, anti–vibration, anti–pressure, anti–drop, water–proof, dust–proof and anti–EMI, etc. The screen has been anti–reflection processed and the design inspiration of appearance roots in "Transformers", bringing joy in vision to consumers.

深圳市怡美工业设计有限公司

Shenzhen IMAY DESIGN Co., Ltd.

咪表运营管理系统

Parking Meter

CDP2007咪表系统的功能按键采用了电容感应技术，一台咪表可同时监测四个车位，并附以高效、节能、环保的太阳能板。此产品采用表面喷粉加工工艺，具有防暴、防水、防涂鸦等特性。

The CDP2007 parking meter is an electronic road-side parking meter with stylish design, outstanding functions and advanced technology. Each can monitor the driving or parking of four cars at the same time through capacitive touch sensing in the function key. In addition, it installs high-efficiency solar panels. The appearance will contribute to construct a harmonious, human-oriented technical platform in the field of static-traffic.

深圳天鹏盛电子有限公司

Keen High Technologies Ltd.

3.5寸闹铃式数码相框

3.5 Alarm Digital Photo Frame

闹铃式数码相框在闹铃基础上添加了数字相框功能，可显示图片；提供2.0音效系统，支持MP3、MP4影片格式，实现了床头全面的多媒体功能。

Alarm Digital Photo Frame works not only as a traditional bed alarm, but a digital photo frame displaying pictures. Besides, it can provide magnified audio system and support mp3 and mp4 format, achieving bedside full multimedia capabilities.

深圳天鹏盛电子有限公司

Keen High Technologies Ltd.

无线便携气象站

Wireless Portable Weather Station

无线便携气象站包含了常见的气象信息，包括天气情况与温度湿度数据，并兼有数码相框的一般功能。屏幕可同时显示天气图标、日期、星期，温度和湿度数据。此外还支持MP3、MP4格式的音频及影像。

Wireless portable weather station can provide you with meteorological information, including weather conditions, temperature and humidity. As a digital photo frame, it can also display pictures, weather icons, date, week, temperature and humidity. It supports MP3, MP4 format as well.

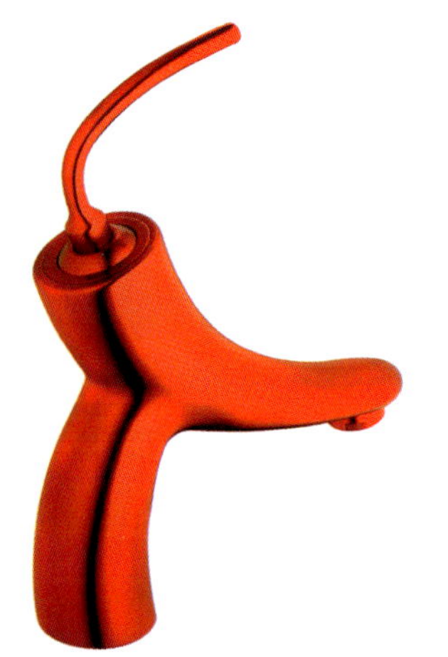

台州嘉德利卫浴有限公司

TAIZHOU CATLY SANITARY WARES Co., Ltd.

卡宴JDL8021000水龙头

CayenneJDL8021000 Faucet

卡宴在西班牙语中有"辣椒"的意思，"辣椒"在中文中意味着"红火"。产品整个造型简单而流畅，表面红色烤漆更是表达出人们的喜悦心情和对美好、红火生活的追求，将美好的愿景巧妙的融合在了外观造型中。

Cayenne means "pepper" in Spanish, and "pepper" indicates "prosperous life" in Chinese. This visually simple product with red baking vanished surface expresses people's longing and pursuit for a booming, happy life.

点石国际有限公司

R.STONE INTERNATIONAL LTD.

照明手套

Lighting Glove

该发光手套照明时双手亦可同时如常工作，还可在掌心掌背随意转动位置，不像一般照明工具手电筒那样需要手持，非常方便。短程光光线，多角度光源，针对日常家居用品市场，满足家中突发停电，尤其进行一些桌上DIY工作的需要。

Lighting Glove, an innovative illumination tool, sets people's hands free while doing handy work in the dark. It also can change the light angle under different situations. For example, light can turn to back side or to the center of the palm. This product is mainly for tiny desktop or DIY job.

盐城工学院

Yancheng Institute of Technology

理想屋——多功能组合移动空间家具

Dream House—Multifunction Moving Unit Space

小型移动空间"理想屋"拆装、运输便捷。屋顶装有半透明卷帘，卷帘放下，"小屋"便成为独立的闭合空间；卷帘卷起，"小屋"又变成半开放式。该产品的出现解决了家具体积大不便于运输等问题。

Dream house is a small moving space which can be disassembled and moved easily. The roof is installed with translucent roller blind. With the blind unfolded, the house becomes an individual close space; with the blind rolling up, it changes to a half-open space. The product gets rid of the difficulty in house moving.

太仓市贤信堂设计咨询有限公司

SINCERE TONE DESIGN AND CONSULTING CO., LTD

卡布&奇诺一家玩具

Furnishing Toy

卡布&奇诺主要面向青年人，既可作为玩具，又是舒适、富有安全感的座椅，靠墙又成了睡垫。两只"耳朵"上装有微型音响可连接外置播放器；"眼部"的香薰带可放置香薰产品；手臂还设有置物袋。

Capu & chino is a combination of toy and furniture, facing to the young generation. The two mini speakers installed on capu & chino can connect to the outboard player. The champignon bag in tits eyes can contain dry flowers and spices. Remote controller or some small things can be put in the arms.

太仓市贤信堂设计咨询有限公司

SINCERE TONE DESIGN AND CONSULTING CO., LTD

时・光——幻彩LED计时器

Colorful LED Timer

彩色光区的随意变幻，是通过程控技术，利用程控芯片中的特殊程序来实现操作。采用LED光源，比普通的液晶屏节约大量能耗。

The change of the colorful lighting area is controlled by remote control technology, which applies the special program of the controller chip. The application of LED can display the timer, and at the same time save more power than the ordinary LCD screen.

威刚科技

ADATA

运动登山碟S805

A-DATA's S805

威刚S805外形简约，不对称扣环挂钩设计独特；以锌合金打造外观质感，采用侧边旋转USB连接头的无笔盖结构；内附2GB~16GB的内存，通过USB 2.0接口储存档案。简约外型，搭配实用性功能，视觉上及使用上均呈现出运动时尚的独特品味。

Featuring mountain-climbing sky hooks, the A-DATA's S805 USB Flash Drive blends delicate craftsmanship in with a touch of sporty style. Embraced by zinc alloy frame, it incorporates swivel design to eliminate cap-losing problems. The karabiner can be easily clipped on a key chain or a handbag. Capacity ranging from 2GB to 16GB and the USB 2.0 high speed transmission can perfectly store your documents.

厦门人水卫浴有限公司

上善若水水龙头

"上善若水"意思是：最高境界的善行就像水的品性一样，泽被万物而不争名利。本产品简洁、干练的线条显示出温和却笃定的感觉。把手上扬，易于操作；按下绿色透明小按钮则能可实现50%的节水。

Xiamen Renshui Industries Co., Ltd.

As Good As Water Faucet

"The highest excellence is like that of water" means: the highest realm of good deeds is like the character of water, which benefits all things without struggle for fame and gain. With the handle lines up, operation becomes much easier. Pressing the green button can save 50% of water which emphasize the concept of resource-saving.

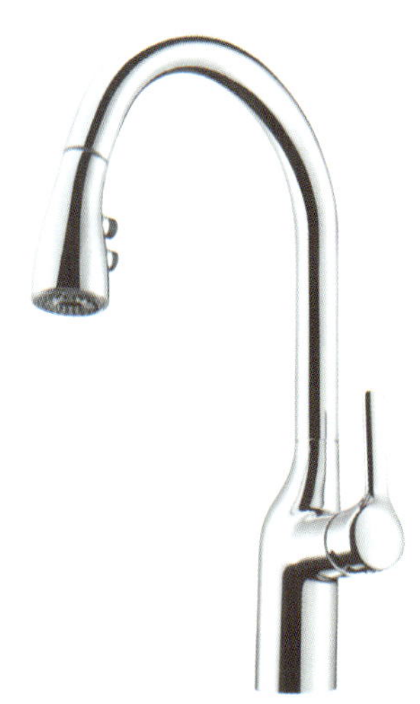

厦门人水卫浴有限公司

光之韵水龙头

本设计利用水力发电的原理，将出水的能量转化为电力。自带感温系统，颜色随水温变化而改变，提醒人水的温度；龙头的头部可自如抽取，多角度冲洗；多功能按钮切换花洒提供多种水花，满足厨房清洁的不同需求。

Xiamen Renshui Industries Co., Ltd.

Shine Faucet

This design applies the principle of hydro-electric power, which transforms the energy from water to electricity. Its features include: self-control temperature system, color changing in accordance with the temperature to remind people of the temperature. The pull-down spray head is super ergonomic and easy-to-use. Multi-function spray option provides users with several flow patterns.

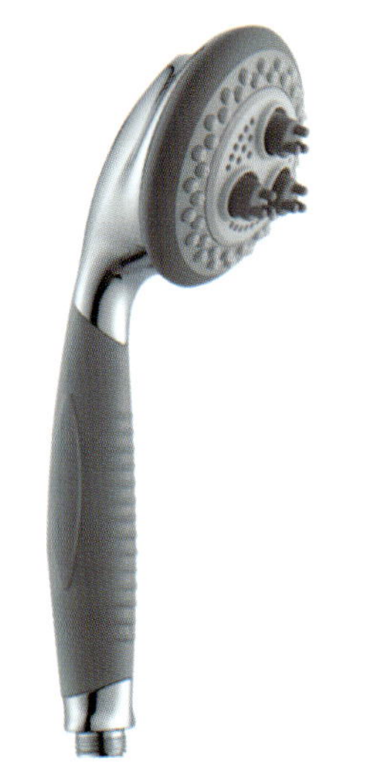

厦门人水卫浴有限公司

悠然花洒

该产品是花洒和按摩器的结合体，通过水压驱动按摩球实现按摩；手柄及按摩颗粒采用软胶设计符合人体工程学原理；造型采用仿生设计手法，选用了鹅的头部形态。

Xiamen Renshui Industries Co., Ltd.

Leisureliness shower

It is a handheld shower as well as a massager, applying hydroelectricity technology on the massage. Adopting ergonomics, the handle and massage particle is wrapped with rubber. Rotating massage head makes user feel more comfortable. Out of bionic design methods, the overall shape looks like a gooseneck.

厦门人水卫浴有限公司

触你所想花洒

触你所想花洒拥有创新功能切换方式。使用者无须通过转动转环切换水花，只需触摸花洒本体，就能享受不同水花所带来轻松舒适的淋浴体验，操作更便捷、富有情趣。

Xiamen Renshui Industries Co., Ltd.

Touch Your Wish shower

Touch Your Wish is a rain shower with the innovative function of switching water flows. Traditional rain showers utilize swivel to select different functions. But this product allows users to control water flows by just touching, providing users with more convenient operation and delectable shower experience.

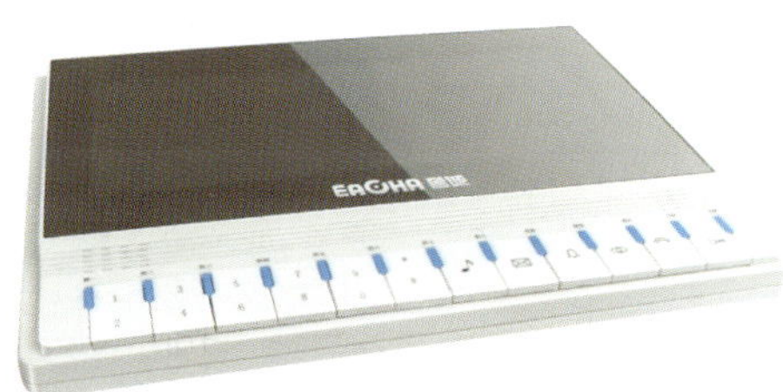

厦门智慷电子科技有限公司

智能家居——室内分机

Pianobell外观上凸显钢琴似的简洁、高雅；功能上在可视门铃基础上增加了家庭数码相框、个性彩铃设置、访者记录等多种功能；表面整体式的橡胶透光指示灯与按键结合，在实现指示功能的同时结构上也有了比较大的创新。

Xiamen Zikam Electronic Science and Technology Co., Ltd.

Building Automation-indoor Phone

Home indoor phone PIANOBELL highlights elegance and nobility in outlook design, and adds a digital photo frame, polyphonic ring tone, and record of visits to the Video Door-phone system. For product surface processing, high brightness effect is adopted to embody the purity and grandeur features of piano; the combination of overall rubber crystal pilot lamp and button optimizes product structure.

显域有限公司

抱抱储存系列

“抱抱”存储系列包括：有连体盖子的杂物储存箱、光盘储存箱、有透明盖顶的被衣储存箱以及尼龙布料、内有夹棉的洗衣篮，使生活空间井井有条又充满了时代感。

Idencity Ltd.

Hug & Hug Storage Series

Hug & Hug Storage Series include Cubic Storage Box with its lid attached, Discs Storage Box, Blanket/Clothing Storage Box with a dust-proof transparent lid and Laundry Hamper made of quilted nylon, offering you a tidy and modern living space.

郭洪生

小水滴电动车

小水滴电动车，全封闭设计，遥控车门，可乘坐三人，造型美观时尚，纯绿色能源产品。联合国复函称此电动车是大众节能减排产品，市场巨大。此车目前广受欢迎，供不应求。

GUO HONG SHENG

Small Waterdrop Solar Energy Electric Car

The characteristics of water-drop electric car are: remote keyless entry, capable of holding three people, fashionable appearance and pure green energy adoption. The United Nations states by letter that this product is energy saving and it will have a huge market. At present, the supply of this appealing electric car falls short of demand.

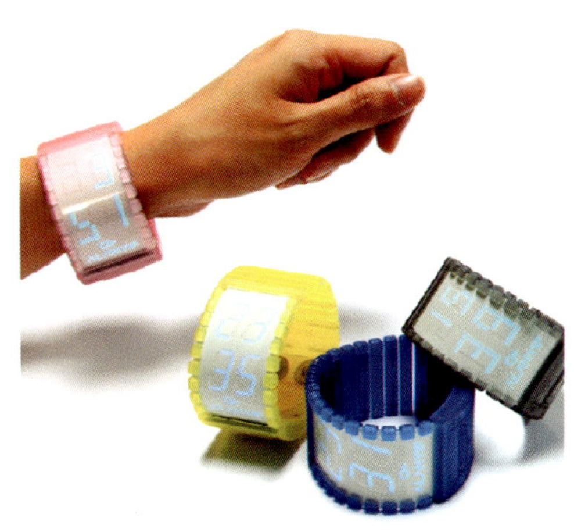

叶智荣工业设计（深圳）有限公司

寿司计时手镯

寿司计时手表糅合最新发光膜技术，既薄（少于0.6mm）又柔软的发光显示屏幕，提供了最佳人体工程学弧度，以贴合不同用户的手腕；高强度磁力接合扣增加了现代高科技佩戴感。

YIPDESIGN.Co., Ltd.

SushiCal Calculator

Sushical Calculator adopts high-end EL technology that provides a thin and flexible light up display (less than 0.6mm). The display fits different wrist of users. The Bracelet employs strong magnetic locks to project a modern and hi-tech feeling during operation process.

易造工业设计（北京）有限公司

掌上书院阅读器

“朴素、自然、纯粹”是该产品设计的关键，具体体现在隆起的书脊、微弧的书页，修长的机身比例，还有和机身融为一体的操作按键上。按键的起伏、切削和机背防滑区，机壳中的音效给用户提供了惬意的阅读体验。

Exmade Design Consultants Co., Ltd.

Easy Reader

"Simple, natural, pure" are three key words in its design, you can find them in raised backbone, pages with arc line, and the buttons integrated into the slender body. The independent operation area, the slip resistance area on the back shell, and the big resonant chamber in 10mm thick offer users a pleasant reading experience.

易造工业设计（北京）有限公司

Exmade Design Consultants Co., Ltd.

宣爱模拟车

Flying Wings

该产品巨大的双翼、向上倾斜的翼尖以及双翼间160°的视场无不体现出设计中"强壮、速度、前卫"的灵魂，给驾驶者一种身临其境的感受。另外，模拟驾驶训练与传统实车驾驶训练相比更省油更环保。

When you see the strong wings and the 160° field of vision created by these expanded wings, you can touch its soul of "strong, speed and avant-garde". Compared with the traditional real driving experience, simulation driving can save more gasoline and is more environmental-friendly.

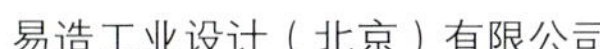

易造工业设计（北京）有限公司

Exmade Design Consultants Co., Ltd.

手持条码扫描终端

Portable Barcode Scanner

手持条码扫描终端定位于专业数据终端，是一款集条码采集、多种无线及有线数据通信、人机交互、电池供电及充电管理、打印、软件及硬件扩展等功能全面的PDA产品。人性化的设计，手持移动、桌面放置均可使用。

Portable Barcode Scanner is positioned in professional data terminal, which is a PDA to gather a punch of functions including bar code collection, many kinds of wireless and wired data communication, human-computer interface, power management, printing and software/hardware expansion. This device is a humane design, which could be used in hand or on the desk.

易造工业设计（北京）有限公司

Exmade Design Consultants Co., Ltd.

"状元、小生、财神"音响

Beijing Opera Idol

"状元、小生、财神"音箱将传统元素融入了现代产品设计中——小巧的造型、全新的发声方式、简单并带有娱乐性的操作，摆脱了以往音箱刻板的形象，成为传承中国文化、凸显现代科技的交点。

"Beijing opera idol" vibration speaker integrates traditional elements with modern trend. Discarding rigid outlook of current speakers, the lovely faces of this product adopting characters in Beijing opera makes the Chinese culture and fashion be known by the world.

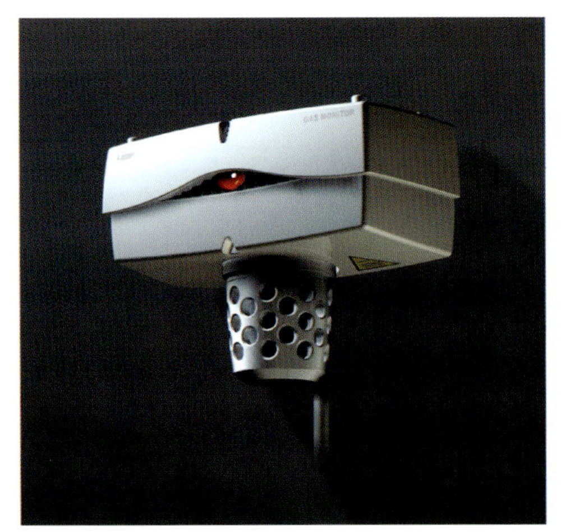

赵斌

ZHAOBIN

危险气体预警系统

Early Warning System for Dangerous Gases

这套系统灵敏、坚固、防爆，不会误报。预警，包含监测和警告的含义，系统中的传感器被设计成具有猎豹般眼神的形态语义，最大程度传达出这套系统灵敏且可靠的性能。

Early warning, monitoring and warning contained in the meaning of the sensor system are designed to look like a cheetah with a form of semantics, to maximize the system to convey a sensitive and reliable performance.

浙江雅鼎卫浴股份有限公司

Zhejiang Yatin Bath Corp.

源远流长面盆龙头

Swing Faucet

这款龙头用曲线生动刻画了水的柔美，又与汉字"水"的写法不谋而合，优雅弧线使龙头的主要设计元素有了无限延伸的可能，更有与其配套的挂件产品，与龙头交相辉映，尽显和谐之美。

This faucet series vividly depicts tenderness of water and harmonizes with the handwriting of ancient Chinese character of "water". Through precise calculation and repetitious tests, the graceful design enables the main element of faucet to extend infinitely. This series faucet matches accessories well, displaying the harmonious beauty.

浙江中克家居用品有限公司

Zhejiang Zhongke Household Supplies Co., Ltd.

水到"曲"成CHUNGO1801水龙头

Water Comes to a Bend Faucet

本产品在外观设计上借鉴河流之形，给人舒展自如之感。出水部位宽阔、平整如河流的平切面带有雍容开阔之意；自然的弧线造型，层次分明；辅之以冷素的水银色，使其呈现梦幻般的奇妙感受。

The outline of the product looks like a winding river, not only shows the feeling of free stretch, but expresses designers' view of maintaining the balance of nature.

震旦办公家具设计中心

AURORA Furniture Co., Ltd. Shanghai

斯巴克沙发

SPARK Sofa

震旦家具SPARK当代型沙发融合了明式椅的神韵和简约的欧式风格，直线、曲线协调运用，温润木纹质感与细腻面料搭配，自然和谐、时尚雅致；"拥抱"符号象征友好待客之道。

The sofa Spark integrates the features of chairs in Ming Dynasty and the concise style of furniture in Europe, employs straight lines and curve lines, wood texture and soft cloth, thus harmonious, fashionable, natural and elegant. The symbol "hug" indicates hospitality.

震旦办公家具设计中心

AURORA Furniture Co., Ltd. Shanghai

明德沙发

MIND Sofa

这是一款适用于主管空间及贵宾会所的接待型沙发，传承中国经典文化，提倡"厚德载物"的领导者风范。精心设置的腰靠，可提供有力腰部支撑。秉明德，载尊贵。整体造型豪迈、大气。

This magnificent sofa is suitable to supervisors' offices and VIP rooms. It features Chinese cultural characteristics and leaders' quality that highlights "social commitment". Carefully designed backrest can perfectly support users' waist.

北京致翔创新产品造型设计有限公司

Bauhaus Aestech Industrial Design Consultancy

lomo相机—公主日记

Lomo Camera - Princess Diary

LOMO公主日记相机，粉红色的立体桃心搭配一对雪白的翅膀的镜头盖成了机身的焦点，同时镜头盖内侧设计成了一个微型相框，可以贴大头照。机身主体材料采用透明软胶材质，晶莹、可爱，而且增强了手感。

LOMO princess diary camera is very stylish. The lens cap is designed into the shape of heart shape with a pair of snow-white wings. The lens cap is also a mini photo frame for full screen callers. The body is made of transparent soft rubber, attractive and beautifully carved.

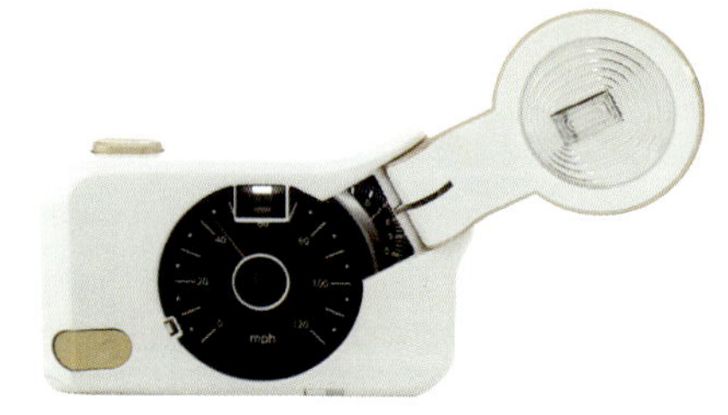

北京致翔创新产品造型设计有限公司

Bauhaus Aestech Industrial Design Consultancy

lomo相机—黄金眼

Lomo Camera-Gloden Eye

该相机按下按钮后，镜头盖向斜上方弹出，露出镜头与巨大的闪光灯；镜头盖与闪光灯的整合使相机操作更容易；构造上沿袭古董相机的特点，弹出的闪光灯与镜头间距尽量拉大，保证良好的成像质量与明暗层次。

When the button of LOMO golden eye camera is pushed, the lens cap is popped out to upside, exposing the lens and the big flashlight; the integrated lens cap and flashlight make the camera easy to use; concerning structure, it heritages the characteristics of antique camera, the popped flashlight is distant enough to the lens, ensuring the high quality of picture and shade of grey.

北京致翔创新产品造型设计有限公司

Bauhaus Aestech Industrial Design Consultancy

整体厨房—精英型

Kitchen-Elite

该厨房主要面向社会精英人士。落地柜搭配厚度仅18cm的超薄吊柜，使用更方便；12cm厚的人造石操作台能充分体现橱柜的专业与品质，异型中岛设计集合酒柜、微波炉、电烤箱、电视等电器，还有充足的存储空间。

This is a kitchen designed for elites. The cabinets with 18cm thick walls, the 12cm thick CORIAN worktops and a drawer inside are more convenient. The attractive textures of wood, aluminum, and glass make the kitchen modern and high-level.

北京致翔创新产品造型设计有限公司

Bauhaus Aestech Industrial Design Consultancy

整体厨房—主妇型

Kitchen-Housewife

主妇型厨房主要体现功能的提升，布局的合理性及外观的柔美。中厨设置在隐藏的高柜中，阻挡了油烟的外漏。中央岛台通过使用人造石、玻璃、金属与木纹几种不同材质的搭配突显品质感。

The kitchen presents a high end functional concept; it has reasonable layout and elegant outlook. To meet the domestic housewives' needs, besides eastern style, western elements are also used in the central part, for the concern of communication need both for the housewives and their families and guests.

中国华录集团有限公司

蓝光播放机

蓝光播放机整机尺寸为430mmx270mmx43mm，为全球首创的超薄机型；前面板搭配倾斜镜片，增加产品视觉冲击力；45度按键符合人体工学，使用更方便更舒适；顶板的蓝光LOGO下沉设计是国内同类产品首创。

CHINA HUALU GROUP CO., LTD.

Blue-ray Disc Player

The dimension of Blue–ray Disc Player is 430mmx230mmx43mm, which is the first ultrathin type in the world. Visual impact is enhanced by adding optic lens to the front panel. More comfortable and convenient operation is realized through ergonomic design of 45 degree keystroke. The subsidence Logo on the top plates/crest slabs is unique in China.

中山华帝燃具股份有限公司

雅韵系列

雅韵系列产品包括烟机、灶台和消毒柜三部分，其中烟机采用华帝独特的自动清洗技术；灶台极佳地满足了中式煎炒烹饪方式的火力要求；消毒柜则呈现给用户一套结合臭氧和紫外线技术的专业化消毒设备。

Vatti Gas Appliance Stock Co., Ltd.

Curve

The Curve series includes a range hood, a cook top and a sterilizer. With the Curve Range Hood, the auto clean technology is a trustable answer to cleaning and hygiene in China. The Curve cook top features the latest Vatti expertise in burner technology. And the sterilizer adopts ozone and ultraviolet technique.

中兴通讯股份有限公司

H200 V1.1家庭网关

此产品可向家庭设备提供Internet网络连接以及数据、语音、视频等多种方式的通信服务。简洁圆润的造型，修饰以有力的线面，增添了产品的个性。使用透明材质罩住机体，减少了人机间的距离感。

ZTE CORPORATION

Home Gateway

This gateway features central control and management of all devices that get access to the Internet through it. It provides various communication services such as Internet access, data, audio and video. The simple and mellow appearance makes it more distinctive.

中兴通讯股份有限公司

ZTE CORPORATION

ZTE H200 下一代家庭网关

ZTE Home Gateway

ZTE中兴 ZXV10 H200下一代家庭网关，运用计算机网络技术和宽带网络接入技术，建立家庭网络中心。创新性的硬朗线条和块面结构，突破了家庭网关简单方盒子概念，是设计的亮点，体现了家庭网关的前瞻性与实用性。

This gateway features central control and management of all devices that get access to Internet through it. It has a powerful form with sharp curves especially for the home life-to-be. Another feature is the stagger-peaked layer of the surface which can keep the gateway cool and work safely.

珠海市润星泰电器有限公司

ZHUHAI RUNXINGTAI ELECTRICAL EQUIPMENT Co., Ltd.

“艾乐家”独创的全自动智能恒温型美容热水器

IWHO-C Series

“艾乐家”独创的全自动智能恒温型美容热水器，是现代美容与科技沐浴的新产品。既可独立使用，又能与热水器组合使用，满足现代人的个性化高品质生活。产品面壳采用IML模内转印工艺,可做个性化DIY设计。

"ORAGER",a unique, fully-automatic, smart, constant temperature water heater, can be used independently or together with a water heater so as to meet diverse needs of users. Adopting IML in-mold decoration, personalized design can be made according to uses' requirements.

珠海市润星泰电器有限公司

ZHUHAI RUNXINGTAI ELECTRICAL EQUIPMENT Co., Ltd.

“艾乐家”智能恒温型旋钮热水器

IWHO-G Series

“艾乐家”智能恒温型旋钮热水器（IWHO-G系列）采用大流量物理防电模块技术，水电隔离技术，发热杯特有的接地技术等多重主动安全保护技术，确保消费者使用安全。产品采用飞梭旋钮开关，数码显示，易于操作。

It is a patented invention for "ASF" (Anti-Scalding Function) to prevent scalding from overheated water. Free-adjusting rotary knob and digital display make it easy to control. Security technologies such as high water volume, leakage protection module, advanced water-electricity separation as well as the special earth-connecting technology for heating cup are adopted so as to ensure the user security.

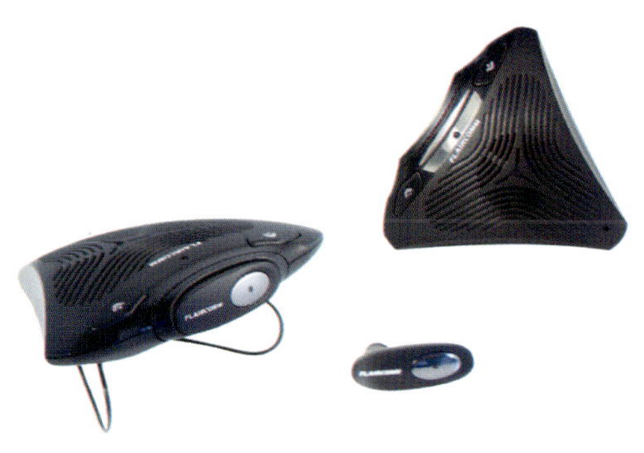

拙雅科技有限公司

BTHF361蓝牙子母车载免提

该产品实现了"户外"蓝牙耳机、"车载"免提底座、"小型电话会议"一机一键式操作，使蓝牙耳机和车载免提在技术、功能、人机关系等的融合设计上，给用户带来了全新使用体验，提升了通话质量和驾车的安全性。

Joya Science&Technology Co., Ltd.

Integrated Car-kit Bluetooth Hands-free Devices

This Car–kit Bluetooth Device innovatively achieved the "one–touch operation" of small telephone conference with outdoor bluetooth headset and car–kit base. It provides a brand new experience for users and hence ensures the safety of driving.

拙雅科技有限公司

军工手机

军工手机具有良好的三防效果，可在水下浸泡后使用。背面增加了一个独立的模块，不但具有气压仪、高度仪、指北针、温度仪的功能，为军人在户外提供实用功能，而且使用独立电源，手机没电后还能继续使用。

Joya Science&Technology Co., Ltd.

Military Mobilc Phonc

This is a water–proof, dust–proof and shake–proof multi–functional mobile phone, which is particularly suitable for military use or outdoor campaigns in harsh environmental conditions. The whole appearance design is composed of smooth lines, fine edges and special green color, which well represents the image of army. And the use of an independent power supply enables you to continue to use the mobile phone after it gets power–off.

拙雅科技有限公司

商务微型投影仪G40

G40，是一款商务微型投影仪，极具便携性，内置电池的使用增强了其便携功能；它的AV和VGA端口适合各种设备的视频链接；在无任何外界电源情况下可以实现视频、图片、数据表格等内容的播放。

Joya Science&Technology Co., Ltd.

Micro Projector

G40 is a micro–projector for business which can be connected to notebooks, mobile phones and other digital products. It meets the needs of modern business and reflects the realization of the portable projector. It is outstanding for its ultra–tiny size. It can still play videos, pictures, data forms and so on without external power supply.

北京江河幕墙股份有限公司

BEIJING JIANGHE CURTAIN WALL Co., Ltd.

新型节能防沙防雨型通风百叶

New Type Energy Saving Anti-sand & Weathering Ventilation Louver

采用三层百叶片设计，具有通风、防雨、防沙的功能。采用铝合金型材，可二次回收。产品设计便于加工和安装，作为自主研发产品，与进口产品相比，每平米节约1万元人民币。

This product carries many functions such as ventilation, rain-proof and sand prevention. It is made of aluminum that can be recycled. It's designed to be fabricated and installed easily which can realize the modernization, standardization of mass production. The rich colors are not only elegant but also match the architectural styles of the building.

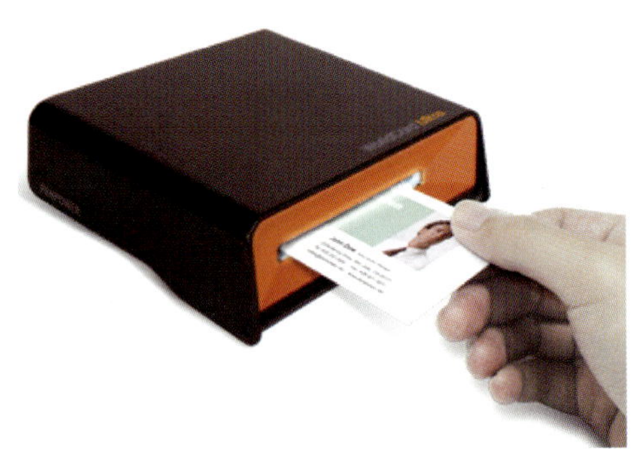

北京蒙恬科技有限公司

Penpower(China) Technology Ltd.

名片王极致版

WorldCard Ultra

方便快速有效的管理名片，自动归到正确字段，并能辨识高达20种语言，能够整合电子邮件及手机等行动装置信息。

As an elegant-palm-sized color business card scanner, WorldCard Ultra takes only 3 seconds, 3 steps to manage business card data effectively. It can recognize more than 20 languages, and integrate with e-mail, cell-phone, PDA and other mobile device information.

北京蒙恬科技有限公司

Penpower(China) Technology Ltd.

迷你扫译笔豪华版

Penpower Mini ScanEYE Professional Version

符合人体工学设计，文件扫描速度每秒达15厘米，扫描立即进入Word、Excel、Powerpoint等文档中，即扫即输入，免用键盘，方便编撰文件、搜集资料。

Penpower mini ScanEYE Professional Version is an input device which helps you copy books fast, as it can scan 800 characters per minute. The ergonomical design makes it easy to use with a document scanning speed of 15 cm/s. All the scanned documents come into Word、Excel、PowerPoint and so on instantly. Multi-state online translation function is also available for quick inquiry and learning.

贝发集团股份有限公司

KA122700环保笔

用可降解、可回收的纸材制作，节能环保。与一般塑料笔或金属笔相比，纸质笔握感更舒适、柔和，缓解手部疲劳。

BEIFA GROUP Co., Ltd.

KA122700 Environment-friendly Paper Pen

This product is made of recyclable materials such as bamboo leaves, rose leaves or corn leaves, which are more environment-friendly and natural. The pen barrel is softer and more comfortable compared with those made of plastics or metal.

创维-RGB电子有限公司

42M10 家庭娱乐终端

该产品是创维推出的家庭娱乐终端产品，整合了coocaa功能。底部的高光曲面给整体设计添加了动感。深色金属质感边框使产品更具节奏感，功能按键融入边框。内发光电源控制按钮表现了科技的神秘气质。

Skyworth

42M10 Entertainment

As a home entertainment product, 42M10 integrates the entertainment functions of Coocaa, which is quite benefit for customers. Thanks to glossy black color for the main body and the unique smile curve of Skyworth, it gives the design an affinity viewing and is much more relevant to the interior environment. The light from the power button adds to its mysteriousness. Treble speaker and woofer provide high-quality listening experience. It boosts the excellent pictures by the full HD LCD panel.

广东亿龙电器股份有限公司

电磁炉CJ-512

该款电磁炉方便耐用且不失美观。它拥有现代风格的外形，由高质量材料制作而成(不锈钢旋钮、橡胶按键和高级黑晶板)。

ETERNAL (GUANGDONG) ELECTRIC HOLDING LTD.

Induction Cooker CJ-512

The induction cooker enables you to enjoy its convenience, durability and elegance. It not only possesses modern and gentle lines, but also is made of high-quality materials(stainless steel knob, soft rubber buttons and high-grade black crystal plate).

广州市红日燃具有限公司

GUANGZHOU REDSUN GAS APPLIANCES Co., Ltd.

D28吸油烟机

D28 Range Hood

该产品是针对中国消费者对高性能、节能、环保产品的消费需求量身打造的。电机输入功率低至180W，最大排风量高达840m^3/h，噪音低于54dB，令厨房有效摆脱油脂、烟尘、烹饪气味的困扰。

The D28 range hood is designed as a response to the growing demand for top-notch performance, environmental protection and energy saving. It has a maximum air extraction of 840 m^3/h while noise level is limited at 54 dB, but its motor only has a rated power of 180 W. Its patented filters have a 3-Tier Structure which can adequately separate fume and grease from air, keeping the kitchen free of grease, smoke and cooking odors quietly and effectively.

宁波欧琳集团

NINGBO OULIN GROUP

A12型家用吸油烟机

A12 Range Hood

一体成型内胆，方便清洗；智清油网过滤油烟，健康环保；人性化的按钮，一目了然，操作更方便。

This product features the following characteristics: forming one liner to facilitate cleaning; three filtrations to filter oil fume, healthy and green; laser engraving button, easy to recognize and more convenient to operate.

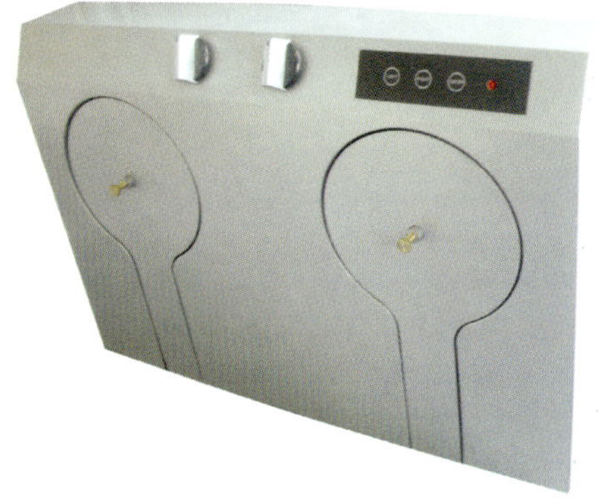

宁波欧琳集团

NINGBO OULIN GROUP

JYL型家用燃气灶

JYL Gas Stove

本产品通过结构革命性创新设计，使家用燃气灶可竖立放置于橱柜台面上，占用橱柜为普通台式或嵌入式灶具占用面积的1/4左右。

Through a revolutionary innovation on structural design, this product enables gas stove be placed on the cabinet stage surface, erect, occupying a small area of the cabinet table, that is, only 1/4 of the space that the current housekeeping gas stoves take.

上海甲秀工业设计有限公司

太阳能计算器鼠标垫

创造性地附加计算器功能，超薄整合设计，并采用第二代高光伏转换率电池板提供电能，节能环保。产品符合人机工学设计，合理设计了计算器键盘的排布位置，便于电脑使用者在使用鼠标或日常办公的时候随时可以使用计算器。

Shanghai TOPRET Industrial Design Co., Ltd.

Calculator Mousepad

Creatively added with a calculator function, the mouse pad is ultra-thin designed. The adoption of the 2-G solar panel with high photovoltaic conversion rate provides its power, which is energy-saving. The design of the product is also in line with ergonomics, as the layout position of the keyboard is properly arranged. Besides, the wrist supporter part is well designed to protect users' wrists. Therefore, users will be able to use the calculator at any time while using the mouse pad or doing daily work.

深圳市中兴移动通信有限公司

GSM手机S610

整机正面采用大面积溅镀工艺，即突出金属高亮的效果，整体色调为金色，进一步提高产品档次。电池盖的设计采用IML工艺，配合正面金属感觉，提升了产品的奢华感。

SHENZHEN ZTE MOBILE TELECOM Co., Ltd.

ZTE-GSM-S610

S610 is a type of low-end bar phone with a 2.0-inch LCD. The highlight of this type of products lies in its small size, dedication, and luxurious feeling. The overall unit adopts large area sputtering process, which highlights the metallic effect. The golden color promotes the product grade on the whole. The battery cover adopts IML process and matches with the metallic feeling of the front case, which promotes the product's luxurious feeling.

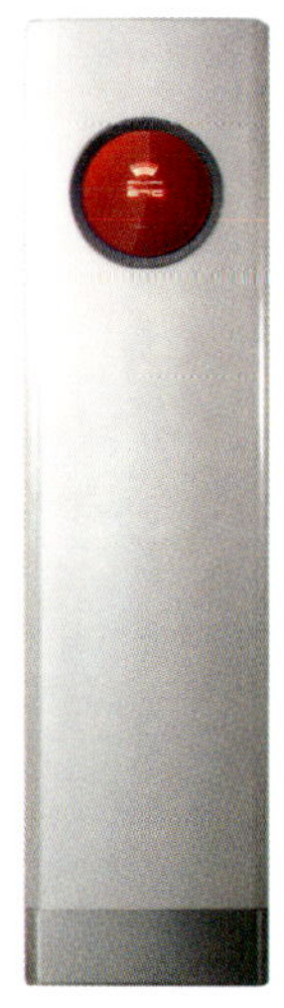

四川长虹电器股份有限公司

长虹Fadin系列空调

操控采用感应按键，即时的声光反馈给予用户良好的操作体验。产品可通过无线网络连入Internet，实现远程监控。将尖端的过滤技术应用于空调过滤系统中，辅以活性炭过滤网，电子集成装置等，有效过滤灰尘及细菌并去除异味。此净化装置在空调关机状态下也能以低功耗独立运行，节能环保。

Sichuan Changhong Electronic Co., Ltd.

Fadin Conditioner

Fadin AC's appearance is pure and natural. The red panel will come out slowly when the Fadin is turned on by touch, blowing strong whirl wind from the two sides of the Air-Conditioner. The room will be cool down or warm up quickly. The red panel is the control interface which also displays air quality, temperature, humidity. The air goes into the product from its back side which prevents the air flow from "short cut", saving energy and working more efficiently. Activated carbon filter is applied to filter dust, germs and smell. And the filter system can work independently in low power-consumption mode, even the Fadin is turned off.

四川长虹电器股份有限公司

长虹DURHI系列等离子电视

显示屏前方环绕着银色边框的部分是扬声器。这种与显示屏分离的设计使得空气能穿过前面板，营造出一种全新的轻盈感。

Sichuan Changhong Electronic Co., Ltd.

DURHI Plasma TV

This slim TV design is featured to express its lightness more than its slimness. The speaker unit lies in the front part with a silver frame. We have left the speaker unit separated from the panel, which proposes the feeling of lightness as the air can come through the front panel.

四川长虹电器股份有限公司

长虹920系列LED电视

设计灵感源于荷花，运用中国传统国画的表现手法——淡墨、玫红色，及大量的留白处理；玫红渐变象征花蕾颜色，融入到产品中，为用户提供一个舒适、高雅的生活环境；按键及软界面设计基于大量的用户研究数据而得，符合人机工学，提升使用的舒适性及便利性。表面采用免喷涂工艺，无污染。

Sichuan Changhong Electronic Co., Ltd.

LED 920

Its design inspiration comes from lotus, using Chinese traditional Chinese Painting's representative techniques, such as light ink, rose color, and blank treatment. Its color ranges from rose to white gradient, just like a blooming lotus, which provides users with comfort and elegant living environment. The design of keys and user interfaces which is based on ergonomics and an amount of statistics from user research can promote usability. Applying paint free technique, it has no pollution to environment.

太阳鸟游艇股份有限公司

80英尺豪华游艇

本设计强调线条含义，以"凤凰"为创意原型，船的窗户外型设计为鸟的羽毛与飞翔羽翼。船的繁杂部件被简化为整体与多功能件以及艺术件，"镜面"新型复合材料与真空成型新工艺的巧妙运用，使船的简洁流畅线条与流光溢彩的块面和谐交融，用以表达现代人追求美好生活、自强不息的"飞翔精神"。

SUNBIRD Co., Ltd.

80Ft. Barge

This product stresses the meaning of lines. Inspired by phoenix, the window adopts the shape of bird feather and wings. The complex parts of the ship are simplified to integral and artistic components. New composite material "Mirror surface" and new vacuum molding technology realizes the harmony integration of lines and surfaces, expressing people's pursuing happy lives and their self-improvement spirit.

北京市三一重机有限公司

SQ170潜孔钻机

集机电液供气于一体，采用履带行走、全液压驱动，高气压潜孔凿岩。适合深孔梯段爆破、深孔预裂爆破、光面爆破、露天矿山削顶及开采、修边护坡等作业。增压空调驾驶室按人机工程学原理设计。"红色之星"涂装，外观新颖，整体感强。双翼式大开门覆盖件，维修保养方便。

Sany Company

SQ170 Down-the-hole Drill

With the integration of gas-mechanics-electronics-hydraulics, it adopts track travelling, hydraulic driving and high pressure drilling. It ideally suits for deep hole blast, the open-air mine, the transport construction and so on. Pressurized air-conditioned cabin designed according to ergonomics to protect the driver. RED STAR paint is novel and gives a sense of completeness. Open coverings are easy to maintain. Hydraulic system adopts load sensitive control system and pilot control system.

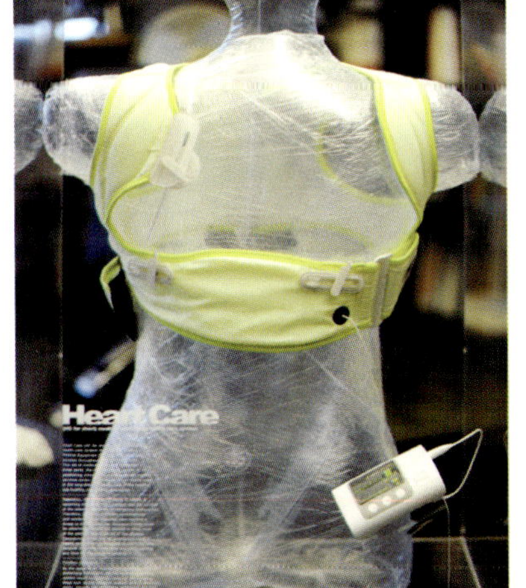

深圳无限空间工业设计有限公司

护心甲

该系统是一套全新的无线全程监护医疗服务系统，由护心甲、护心定位仪、远程护心医疗服务中心三部分组成，广泛适用于健康、亚健康、非健康人群的远程健康监护和预警。

SHENZHEN ND INDUSTRIAL DESIGN Co., Ltd.

Heart Care

Heart Care (HC for short) medical positioning health care system is a set of new wireless throughout monitoring services, composed of three parts: the HC monitoring system, HC positioning instrument, and the remote medical services center. The Product can be widely used in the long-distance health care of the healthy, sub-healthy, non-healthy population.

深圳航嘉创源科技有限公司

魅影H920机箱

一改DIY机箱体积笨重只能立式使用的观念，设计出体积小巧、方便使用的机箱。H920可以卧置用于客厅，连接AV以及大屏幕液晶电视使用，可以组建数字家庭影院，还可以立式用于书房。魅影H920具有百变功能，轻松应对不同家庭需求。

Huntkey Enterprise Group

Huntkey H920 Case

The small, exquisite and convenient DIY chassis is designed to challenge the traditional concept that DIY chassis can only be used vertically. H920 can be used in the sitting room or in the study with a connection to AV and large-screen LCD TV, thus meeting different family demands. It is small-sized, only 12 liters; hard drives, optical drives, graphics cards, side panel all adopt screw-free design; USB 2.0; high-fidelity audio; supporting SD, MMC card reader.

汲运

金漆镶嵌多功能水族大套鼓

该大套鼓外胎为纯木质手工打造，并使用金丝镶嵌工艺雕刻而成。鼓内水域养鱼时无需换水、给氧。如遇停电，鼓内水位自动下降，产生溶氧空间，来电后水可自动充满。内有生化系统和音响功放环绕立体声系统。设计理念新颖实用，使用安全、简洁、节能。

JIYUN

Aquarium Drum

The outer cover of this set of drum is made of pure wood all by hand, and adopts carved gold mosaics. While keeping fish within the drum water, there is no need to change water or to supply oxygen. In case of power failure, the water level inside the drum would low down automatically to produce oxygen space. Inside there are biochemical systems and surrounding sound stereo system amplifier. The design concept is novel and practical, while the application is safe, simple and energy-saving.

汉王科技股份有限公司

创艺掌门

创艺大师三代数位绘图板是一种坐标输入装置，能将操作者的绘画轨迹输入计算机，并识别绘画压力与倾斜角度，再配合相应绘画软件来完成绘画功能。数位板边缘采用弧形设计，有利于手腕的摆放和支撑，更加舒适。绘图区域采用宽屏设计，功能区包含触控环，可实现缩放、滚动、调整笔刷大小、画布旋转等功能。

Hanwang Technology Co., Ltd.

Art Master

The third generation of Art Master digital drawing board is a type of coordinate input device, which can input the operator's drawing trajectory into the computer, identify the pressure and tilt angle when painting and complete the painting in coordination with appropriate software. The edge of the drawing board resorts a curved design, which makes it more convenient to paint with both hands. The product uses a wide-screen design in the drawing area, while the function area use a circular touch ring. Sliding the finger-sensitive Touch Ring can realize the freedom of switching operations such as zoom, scroll, adjust the brush size, canvas rotation.

北京兆维自助服务设备技术有限公司

BEIJING C&W Self-service Equipment Technology Co., Ltd.

场馆自助售票系统

Self-service Ticket Booth System

本场馆自助售票设备采用触摸屏操作模式；支持纸币和银联卡两种付费方式；图文并茂的操作提示贴膜和人工帮助旋钮极具人性化。

This venue ticketing equipment adopts touch screen mode of operation and supports two payment methods—paper currency and UnionPay cards; it can print a variety of formats votes in a single support bar code printing; voice prompts make the operation simple and convenient; equipments at the top of the LED screen can display a variety of information, enabling the staff to operate easily and understand the equipment running state; operation tips with both words and illustrations humanize the operation process.

海尔集团海高设计制造有限公司

Haier Group Haigao Design & Manufacturing Co., Ltd.

海尔i7笔记本

Haicr i7

以产品易用为设计点切入，采用悬浮式键盘缓解手指和腕部疲劳；隐藏式光线感应器能自动调整高亮液晶屏亮度；两侧的线缆接口方便插拔，并预装了针对商务人群的安全软件。生产中参照了环保组织ChemSec和Clean Production Action的绿色生产要求。

It aims at massive business market with thirty million units. The designers started concept from ease of use: The floating keyboard can relieve strain and fatigue; hidden light sensor can automatically adjust the luminance of high contrast LCD display; all slot and connector are placed in two sides which make it simple to remove and insert. The production is complied with Green manufacturing procedure of ChemSec and Clean Production Action.

2009中国创新设计红星奖获奖名单

至尊金奖

产品	单位	页码
九宫盲人手机	中国电信股份有限公司上海研究院	023

金奖

产品	单位	页码
空气净化时尚型	康佳集团生活电器	030
ORIGIN–多功能城市防灾应急微型勤务车	太仓市贤信堂设计咨询有限公司	032
“睿智”–ZTE中兴T8000高端路由器	中兴通讯股份有限公司	034
飞利浦–我的阅读灯	上海木马工业产品有限公司	036
火鹰红外热成像仪	北京蓝威科瑞科技有限公司	038
耳朵沙发	北京曲美家具集团有限公司	040
ONYX电子阅读器	深圳市浪尖工业产品造型设计有限公司	042
专业耳麦PT301	北京品物堂产品设计有限公司	044

最具创意奖

产品	单位	页码
GSR 迷你专业螺丝电起子	齐思工业设计咨询（上海）有限公司	048
香料娃娃(油醋瓶)	桔思创意设计	050
体重指数尺/心衰尺	金唐立鑫（北京）科技发展有限公司	052
彩色电动剃须刀	上海龙域设计	054
汉王电纸书	洛可可工业设计公司	056
禅·静水龙头	厦门人水卫浴有限公司	058
柠檬挤	桔思创意设计	060
模块化可调光度角LED灯具	北京博蓝士科技有限公司	062
无线上网卡AC2766	深圳市中兴移动通信有限公司	064
便携式“皇家普洱”	北京故宫宫苑文化发展有限公司	066

最佳团队奖

获奖者	页码
中国电信上海研究院创新团队	088

最佳新人奖

获奖者	页码
羊文军	090

红星奖

产品	单位	页码
Purity青花酒瓶	DHOME陶瓷产品设计工作室	094
H&D视力表钟	DHOME陶瓷产品设计工作室	094
M50高精磨床	北京博蓝士科技有限公司	094
TCL X10液晶电视	TCL多媒体科技控股有限公司	095
TCL C10液晶电视	TCL多媒体科技控股有限公司	095
爱特那卫浴套装	齐思工业设计咨询（上海）有限公司	095
GBM 6RE, 10RE, 13 RE专业电钻系列	齐思工业设计咨询（上海）有限公司	096
GWS 7–100专业角磨机	齐思工业设计咨询（上海）有限公司	096
GDR 14.4伏/18伏 锂电池专业电钻	齐思工业设计咨询（上海）有限公司	096
中央暖通控制系统	齐思工业设计咨询（上海）有限公司	097
SY4500沥青路面热再生重铺机组	鞍山森远路桥股份有限公司	097
运动支撑内衣AM62311–11	北京爱慕内衣有限公司	097
T1988宽屏液晶电视	北京格物创道科技发明有限公司	098
圆雕《清·乾隆皇帝阅兵像》	北京故宫宫苑文化发展有限公司	098
电子书R6	北京华旗资讯数码科技有限公司	098
数码录音笔(专业型)	北京华旗资讯数码科技有限公司	099

Winner List of 2009 China Red Star Design Award

Best of the Gold Prize

9 Spots "Mobile Global Eye"	Shanghai Research Institute of China Telecom Corporation Limited	023

Gold Prize

Air Cleaner	KONKA GROUP CO.,LTD.	030
Mini Multifunction Emergency Vehicle	SINCERE TONE DESIGN AND CONSULTING CO.,LTD	032
"Sagacity" – ZTE T8000 High-end Router	ZTE CORPORATION	034
My Reading Light	Shanghai Moma Industrial Product Design Co., Ltd.	036
Firehawk	Beijing Lever Create Co.,Ltd.	038
Ear Sofa	BEIJING QUMEI FURNITURE GROUP CORP., LTD.	040
ONYX BOOX	SHENZHEN ARTOP Industrial Design Co.,Ltd.	042
Professional Earphone	PER Design	044

Best Originality Prize

GSR ProDrive Professional	TEAMS Design Consulting Co.,Ltd.	048
KeruKeru	U.I.Design Co.,Ltd.	050
Weight Index Slipstick / Cardiac	JTLX BEIJING Co.,Ltd.	052
Colorful Electric Shaver	LOE DESIGN	054
Hanvon E.pbooks	LKK Industrial Design Co.,Ltd.	056
Zen • Quiet Faucet	Xiamen Renshui Industries Co.,Ltd.	058
LemonRun	U.I.Design Co.,Ltd.	060
"Airflow"—Modular LED Light With Adjustable Beam Angle	IDCdesign	062
ZTE-AC2766	SHENZHEN ZTE MOBILE TELECOM CO.,LTD.	064
Royal Pu'er Tea in Square Pieces	Beijing Imperial Court Cultural Development Company Ltd.	066

Best Team Prize

Shanghai Research Institute of China Telecom Co., Ltd. Innovation Team	088

Best New Designer Prize

Wenjun Yang	090

Excellent Prize

Blue and White Bottle	DHOME CERAMIC PRODUCT DESIGN WORKSHOP	094
Test-Chart Clock	DHOME CERAMIC PRODUCT DESIGN WORKSHOP	094
M50 High Precision Grinding Machine	IDCdesign	094
TCL X10 LCD TV	TCL Multimedia Technology Holdings Ltd.	095
TCL C10 LCD TV	TCL Multimedia Technology Holdings Ltd.	095
Etna	TEAMS Design Consulting Co.,Ltd.	095
GBM 6RE, 10RE, 13 RE Professional	TEAMS Design Consulting Co.,Ltd.	096
GWS 7-100 Professional	TEAMS Design Consulting Co.,Ltd.	096
GDR 14.4V-Li, 18V-Li Professional	TEAMS Design Consulting Co.,Ltd.	096
Climatix	TEAMS Design Consulting Co.,Ltd.	097
SY4500 Asphalt Road Surface Hot-regenerating Re-paving Unit	AnShan SenYuan Road and Bridge Co.,Ltd.	097
Sport Corsage	Beijing Aimer Lingerie Co.,Ltd.	097
T1988 LCD TV	BEIJING CREIWAY INVENT&DESIGN CO.,LTD.	098
Emperor Qianlong Inspecting The Troops	Beijing Imperial Court Cultural Development Company Ltd.	098
E-book R6	Beijing Huaqi Information Digital Technology Co.,Ltd.	098
High-End Digital Voice Recorder	Beijing Huaqi Information Digital Technology Co.,Ltd.	099

红星奖

产品	企业	页码
方正VEFOUND "文房" 3G电子书	北京华新意创工业设计有限公司	099
双层智能呼吸窗1系	北京嘉寓门窗幕墙股份有限公司	099
御米油(罂粟籽油)系列包装	北京蓝海洋文化传播有限公司	100
猎鹰监控系统	北京蓝威科瑞科技有限公司	100
狩猎者红外夜视镜	北京蓝威科瑞科技有限公司	100
桌面3G路由G3R01	北京品物堂产品设计有限公司	101
OBU电子高速计费器	北京品物堂产品设计有限公司	101
家庭无线网关G3R09	北京品物堂产品设计有限公司	101
无线多媒体信息交互终端机	北京品物堂产品设计有限公司	102
电子阅读器Walkreader	北京品物堂产品设计有限公司	102
钥匙扣 CKEYS	北京品物堂产品设计有限公司	102
便携式爆炸物检查仪	北京品物堂产品设计有限公司	103
个人基因分析系统	北京品物堂产品设计有限公司	103
液相色谱仪AJ7302	北京品物堂产品设计有限公司	103
C3椅	北京曲美家具集团有限公司	104
高尔夫实球击打辅助训练系统	北京水童星科技发展有限公司	104
明斯克40升背包	北京探路者户外用品股份有限公司	104
尼尔徒步鞋	北京探路者户外用品股份有限公司	105
帝力可加热冲锋衣	北京探路者户外用品股份有限公司	105
马里可加热冲锋裤	北京探路者户外用品股份有限公司	105
突破防水系列插座	北京突破电气有限公司	106
突破易拉宝插头	北京突破电气有限公司	106
突破旋转变形插座	北京突破电气有限公司	106
园林取土器	北京心觉工业设计有限责任公司	107
ZH5120D立式钻削加工中心	北京信息科技大学机电工程学院	107
无影灯LED OL9570/50	北京谊安医疗系统股份有限公司	107
3in1数字无线话筒	成都意町工业产品设计有限公司	108
四方儿童游戏手柄	成都意町工业产品设计有限公司	108
骨传导耳塞	成都意町工业产品设计有限公司	108
埃森普特半自动弧焊机	成都意町工业产品设计有限公司	109
长江电动喷涂机	成都意町工业产品设计有限公司	109
儿童综合素质测试仪	创意工场（北京）工业设计有限公司	109

产品	企业	页码
银河新星无创血糖检测仪	递加（北京）设计顾问有限公司	110
Podium USB扩充器	广东华南工业设计院	110
LP7000航天按摩椅	东莞市生命动力按摩器材有限公司	110
电脑一体机	多达创新（北京）科技有限公司	111
外置吸入式光驱	多达创新（北京）科技有限公司	111
电源适配器	多达创新（北京）科技有限公司	111
雍容橱柜	方太柏厨设计中心	112
LED数字钟	福州瑞达电子有限公司	112
旋转镜子数显钟	福州瑞达电子有限公司	112
EM1340 "田" 字型投影钟	福州宜美电子有限公司	113
EM711可旋转式太阳能温度计	福州宜美电子有限公司	113
互拷器A2B	高国兴	113
老年人服务平台终端	光彩无限创意中心	114
起动机M105R3040SE	光彩无限创意中心	114
鱼竿—碧波2S	光彩无限创意中心	114
加湿器S30U–H	广东美的环境电器事业部	115
DIY积木概念灯	广东时尚元素生活用品有限公司	115
即热式水壶	广东新宝电器股份有限公司	115
多功能早餐机	广东新宝电器股份有限公司	116
V09系列儿童互动电视	海信集团	116
自由舰橱柜	杭州德意厨具有限公司	116
侧吸式油烟机5360	杭州老板电器股份有限公司	117
聪明套装	杭州老板电器股份有限公司	117
美的凡帝罗三门冰箱	浩汉工业产品设计（上海）限公司	117
长虹饮洞JL7913500水龙头	鹤山市洁丽实业有限公司	118
消防安全管理及监督系统信息终端	恒创碧思特工业设计（北京）有限公司	118
LED工作台灯	黄治中	118
香料娃娃	桔思创意设计	119
沙拉碗	桔思创意设计	119
点心碗	桔思创意设计	119
V12手机	康佳集团	120
9900C手机	康佳集团	120

Excellent Prize

红星奖

Excellent Prize

红星奖

产品	企业	页码
"艾乐家"智能恒温型旋钮热水器	珠海市润星泰电器有限公司	142
BTHF361蓝牙子母车载免提	拙雅科技有限公司	143
军工手机	拙雅科技有限公司	143
商务微型投影仪G40	拙雅科技有限公司	143
新型节能防沙防雨型通风百叶	北京江河幕墙股份有限公司	144
名片王极致版	北京蒙恬科技有限公司	144
迷你扫译笔豪华版	北京蒙恬科技有限公司	144
KA122700环保笔	贝发集团股份有限公司	145
42M10 家庭娱乐终端	创维-RGB电子有限公司	145
电磁炉CJ-512	广东亿龙电器股份有限公司	145
D28吸油烟机	广州市红日燃具有限公司	146
A12型家用吸油烟机	宁波欧琳集团	146
JYL型家用燃气灶	宁波欧琳集团	146
太阳能计算器鼠标垫	上海甲秀工业设计有限公司	147
GSM手机S610	深圳市中兴移动通信有限公司	147
长虹Fadin系列空调	四川长虹电器股份有限公司	147
长虹DURHI系列等离子电视	四川长虹电器股份有限公司	148
长虹920系列LED电视	四川长虹电器股份有限公司	148
80英尺豪华游艇	太阳鸟游艇股份有限公司	148
SQ170潜孔钻机	北京市三一重机有限公司	149
护心甲	深圳无限空间工业设计有限公司	149
魅影H920机箱	深圳航嘉创源科技有限公司	149
金漆镶嵌多功能水族大套鼓	汲运	150
创艺掌门	汉王科技股份有限公司	150
场馆自助售票系统	北京兆维自助服务设备技术有限公司	151
海尔i7笔记本	海尔集团海高设计制造有限公司	151

国庆60周年特别奖

最佳策划奖

作品	页码
星星火炬彩车	070
科技创新彩车	070
文化成就彩车	070
天津——滨海新貌	071
山西——魅力山西彩车	071
上海——腾飞上海彩车	071
浙江——钱江明珠彩车	071
重庆——三峡放歌彩车	072
青海——大美青海彩车	072
香港——紫荆盛放彩车	072

最佳创意奖

作品	页码
农业成就彩车	073
神舟飞天彩车	073
我的中国心彩车	073
北京——北京之歌彩车	074
江苏——吉祥如意彩车	074
江西——崛起江西彩车	074
山东——岱青海蓝彩车	074
河南——花开盛世彩车	075
广东——领潮争先彩车	075
贵州——多彩贵州彩车	075

最佳技术奖

作品	页码
牡丹花道具彩车	076
社会主义新农村彩车	076
交通成就彩车	076
生态环保彩车	077
和谐社区彩车	077
辽宁——振兴乐章彩车	077
福建——扬帆海西彩车	077
湖北——凤舞楚天彩车	078
台湾——宝岛台湾彩车	078
绘就蓝图彩车	078

最佳视觉奖

作品	页码
浴血奋斗彩车	079
艰苦创业彩车	079
体育成就彩车	079
北京奥运彩车	080
众志成城彩车	080
黑龙江——龙腾盛世彩车	080
广西——壮乡欢歌彩车	080
西藏——和谐西藏彩车	081
甘肃——盛世华章彩车	081
宁夏——塞上江南彩车	081

最佳造型奖

作品	页码
能源成就彩车	082
教育成就彩车	082
同一个世界彩车	082
团结奋进彩车	083
内蒙古——草原飞虹彩车	083
吉林——精彩吉林彩车	083
安徽——江淮和畅彩车	083
湖南——锦绣潇湘彩车	084
云南——七彩云南彩车	084
陕西——三秦新韵彩车	084

最佳制作奖

作品	页码
工业成就彩车	085
民主政治彩车	085
依法治国彩车	085
人口卫生彩车	086
河北——激情河北彩车	086
海南——绿色海南彩车	086
四川——奋进四川彩车	086
新疆——天山祝福彩车	087
澳门——盛世莲花彩车	087
繁花似锦彩车	087

Excellent Prize

Special Prize for the 60th Anniversary of the Founding of PRC

Best Planning Award

Best Originality Award

Best Technology Award

Best Vision Award

Best Model Award

Best Execution Award

红星照耀中国 是创新的中国

Red Star over China – Innovative China